SERIES EDITORS:
Stewart R. Clegg &
Ralph Stablein

Bob Hodge, Gabriela Coronado,
Fernanda Duarte and Greg Teal

Chaos theory and the Larrikin Principle

Working with organisations in a Neo-Liberal world

ADVANCES IN ORGANIZATION STUDIES

Liber
Copenhagen Business School Press
Universitetsforlaget

Chaos theory and the Larrikin Principle

ISBN 978-91-47-09484-4 (Sweden)
ISBN 978-82-15-01737-2 (Norway)
ISBN 978-87-630-0235-6 (Rest of the world)
ISSN 1566-1075

© 2010 Bob Hodge, Gabriela Coronado, Fernanda Duarte, Greg Teal and Liber AB

Publisher's editor: Ola Håkansson
Series editor: Stewart Clegg and Ralph Stablein
Typeset: LundaText AB

1:1

Printing: Sahara Printing, Egypten 2010

Distribution:
Sweden
Liber AB
S-205 10 Malmö, Sweden
tel +46 40-25 86 00, fax +46 40-97 05 50
http://www.liber.se
Kundservice tel +46 8-690 93 30, fax +46 8-690 93 01

Norway
Universitetsforlaget AS
Postboks 508
NO-0105 Oslo
phone: +47 14 75 00, fax: +47 24 14 75 01
post@universitetsforlaget.no www.universitetsforlaget.no

Denmark
DBK Logistics, Mimersvej 4
DK-4600 Koege, Denmark
phone: +45 3269 7788, fax: +45 3269 7789
www.cbspress.dk

North America
International Specialized Book Services
920 NE 58th Ave., Suite 300
Portland, OR 97213, USA
phone: +1 800 944 6190
fax: +1 503 280 8832

Rest of the World
Marston Book Services, P.O. Box 269
Abingdon, Oxfordshire, OX14 4YN, UK
phone: +44 (0) 1235 465500, fax: +44 (0) 1235 465555
E-mail Direct Customers: direct.order@marston.co.uk
E-mail Booksellers: trade.order@marston.co.uk

⚠ **All rights reserved**
No part of this publication may be reproduced, stored in a retrival system,
or transmitted, in any form or by any means, electronic, mechanical, photo-copying,
recording, or otherwise, without the prior written permission of the publisher.

Advances in Organization Studies

Series Editors:
  Stewart Clegg
  Professor, University of Technology, Sydney, Australia

  Ralph E. Stablein
  Professor, University of Otago, New Zealand

Advances in Organization Studies is a channel for cutting edge theoretical and empirical works of high quality, that contributes to the field of organizational studies. The series welcomes thought-provoking ideas, new perspectives and neglected topics from researchers within a wide range of disciplines and geographical locations.

www.organizationstudies.org

Contents

Acknowledgements 9
Preface 10

CHAPTER 1
What's wrong with business education? 29
1. The crisis in business education 29
    Outline of a critique 30
    The Management Education machine 33
    Resistance is futile – isn't it? 35
2. Towards a critical pedagogy 37
    Management vocabulary and its taboos 38
    Tensions in the Managerial Complex 39
    Critical pedagogy and management studies 41
3. Learning from where you are 44
    What are students really like? 44
    Three-body Critical Pedagogy 46
    What should management teachers know? 49
    The scope of 'Management' 50
    Making sense of the big picture 52
    The crisis in business education 53

CHAPTER 2
The Larrikin Principle 54
1. Larrikin practices 55
    Defining the larrikin 56
    The double face of larrikinism 58
    Larrikins, larrikinas and the gender problem 60
2. Fuzzy analysis and the Larrikin Principle 61
    Larrikin myths and meanings 65
    Uses of history 67
    The colonial factor 69

    Larrikin politics: Is Rudd a larrikin?.................................... 70
    Cybernetics of the larrikin.. 71
3. Larrikins in management eyes.. 73
    Whistleblowers... 76
    The Larrikin Principle... 78

CHAPTER 3
Managerialism 79
1. Larrikin perspectives on management................................. 80
    Larrikin diagnoses of Managerialism................................. 82
2. The double face of Managerialism...................................... 86
    Total Quality Management as paradigm shift?.................... 87
    Taylorism revisited.. 89
    Taylorism and the managerial ideological complex............ 90
    Taylor as butterfly effect... 91
    Larrikin antibodies.. 92
    The Star Wars guide to bad management........................... 94
3. Coping with chaos and change... 95
    Discourses of change... 95
    Resistance.. 98
    'Only the paranoid survive'.. 100
    Managerialism... 102

CHAPTER 4
Neo-Liberalism and its disconnects 103
1. The many faces of Neo-Liberalism...................................... 104
    Living with Neo-Liberalism in the developing world........... 106
    Neo-Liberalism down-under.. 108
    Contradictions and the ideological complex....................... 110
    Fractals of the Neo-Liberal complex.................................. 112
2. Globalisation... 114
    Gurus on globalisation.. 116

3. Chaos and freedom **118**
Scientific models of globalisation 121
The globalisation of resistance 123
Alternative media as David vs Goliath........................... 125
The future of Neo-Liberalism 128

CHAPTER 5
Corruption **130**
1. Fuzzy corruption and the fuzzy larrikin **131**
Ambivalent larrikins in business.................................. 133
2. Pseudo-corruption and the Neo-Liberal complex **135**
Wars against corruption... 136
Reporting third world corruption 139
Measuring corruption... 140
3. Discursive corruption.. **144**
The cybernetics of spin... 145
Money as a discursive system..................................... 147
Grand corruption in Larrikin land................................ 150
Deep corruption, Mexican style 151
The consequences of corruption.................................. 153
Corruption 156

CHAPTER 6
Power versus goodness @ the edge of chaos................ **157**
1. The problems of goodness **158**
Globality... 159
Globality at work... 161
Instrumental reason in a chaotic world....................... 164
2. Pathologies of power.. **165**
Pathologies of trust.. 166
Power and control... 168
3. Thriving in chaos **171**
Three body systems and effective partnerships........... 171

    Corporate Social Responsibility and the stakeholder model........ *173*
    Thriving in chaos *176*
    The power of goodness *177*

CHAPTER 7
Soft Capital and the Informal Polity **179**
1. Larrikin correctives to dysfunctional systems **181**
    Brokers for broken systems *181*
    Soft research for soft systems *182*
    The power of charm *184*
    Tricks and dodges *187*
2. Social capital **188**
    Versions of social capital *188*
    Marx on the 'capital' in Social Capital *191*
    Social capital and the cash nexus *195*
3. The informal polity **197**
    Management texts on informal networks *197*
    Informal systems and the idea of social capital *199*
    The governance paradox *201*
    Soft capital and the informal polity *203*

CHAPTER 8
Culture and organisations in a global world **204**
1. The many faces of culture **205**
    Culture in organisations *205*
    Uses of culture and the management paradox *207*
    Culture and globalisation *209*
    What is 'culture'? *210*
2. Ideological complexes and a multicultural world **212**
    Culture and contradiction *213*
    The managerialist ideological complex in cross-cultural
    management *215*
    Management's 'science' of culture *218*

3. Cultural Analysis and the Larrikin Principle 221
Critical discourse analysis and ideological complexes 221
Thick description 223
High art and the meaning of culture 225
Cultural analysis and the larrikin myth 226

Conclusion: a larrikin(a) critique 229
Body 1: the Larrikin(a) Principle 229
Body 2: Managerialism 230
Third Body: Neo-Liberalism 231

Bibliography 233

Subject index 240

Author index 242

Acknowledgements

We firstly thank the Australian Council for Research, and its Humanities panel, for providing the funding for the research on which the book is based. Different versions of the ideas in this book were first presented to constructive and critical audiences in conferences organised by AILASA (the Association of Iberian and Latin American Studies of Australasia) and APROS (the Association for the Pacific Region in Organisation Studies). Among those who contributed stimulating ideas and comments we would like to thank Stuart Clegg and Michael Muetzelfeld.

The University of Western Sydney paid our salaries and provided many other forms of support, plus priceless data. Colleagues in its School of Management provided valuable input: Wayne Fallon, Louise Kippist, Ron Kelly, Patrick McGirr, as did Zöe Sofoulis with generous abundance, plus Wayne Peake and Maree O'Neill at the Centre for Cultural Research.

With this book, some of the family members of the authors provided even more than the love and support which makes larrikin and academic life worthwhile. Geoff Beache wove together management expertise and a larrikin perspective, along with a high level of film criticism. Anna Notley combined a Masters Degree in Accounting with a creative, anarchic streak that made her the ideal reader that we tried to address. Rob Notley lived his complex identity as an exemplary and non-stereotypical larrikin. John Hodge provided a continual critical commentary from the left of our left. Lorena and Mariana Caballero pointed to our naïve idealism which sometimes blinded us to the dark side of breaking rules. We take the opportunity here, as well as later, to thank our three research assistants, Sandra Klinke, Laura Calderon de la Barca, and Beatriz Cardona, who contributed as much to the book as to the life of the project. And we especially wish to express our deep appreciation to all the Australians, Mexicans and Brazilians who contributed with their stories. We thank them all, and add the usual disclaimer: all remaining faults are our responsibility.

Preface

This is a book on management and organisations today, their theories and practices, written as the global financial crisis of 2008 is still unfolding. The crisis is not our theme, and we do not predict how it will all finish, in the short- or long-term. We only want to look at some key issues as they now appear in what some may call the shadow of the crisis. The crisis damaged a dominant edifice in the ideological skyline, which we call, for the moment, Neo-Liberalism. Even if that edifice is rebuilt, to some degree, it has currently lost several storeys or more. Doubts are cast about how solid its foundations were.

In the space it leaves, the light shines through. We can see some things more easily that have been there for a long time, but were minimised or ignored by movers and shakers in business and politics. In this book we ask: which critiques of current practices in management come back with greater credibility and force in this new light? Which alternative ways of thinking about organisations look better and stronger now than they did five years ago, when we began our research? What new possibilities can be glimpsed and pursued in this new situation?

In the USA, President George W Bush was more identified with Neo-Liberalism than any other leader at the time. However, in November 2008, he reacted to the crisis with massive government interventions widely seen as repudiating Neo-Liberal principles. His successor, President Barack Obama, was elected with a mandate to dismantle the policy edifice of Neo-Liberalism, and search for a new basis for national and world governance. In Australia, Prime Minister Kevin Rudd, elected in 2007 from the Centre-Left, wrote a critique of Neo-Liberalism. He called it 'that particular brand of free-market fundamentalism, extreme capitalism and excessive greed which became the economic orthodoxy of our time' (2009:20).

Rudd saw the crisis as a turning point, an event of 'truly seismic significance':

> This is a crisis spreading across a broad front: it is a financial crisis which has become a general economic crisis; which is becoming an employment crisis; and which has in many countries produced a social crisis and in turn a political crisis... It is a crisis which is simultaneously individual, national and global. It is a crisis of both the developed and the developing world. It is a crisis which is at once institutional, intellectual and ideological. It has called into question the prevailing neo-liberal orthodoxy of the past 30 years. (2009:20)

Politics is commonly seen as a distinct branch of life from organisation and management studies. However, leaders like Bush, Obama and Rudd are all in effect CEOs of huge, highly complex organisations. All three draw on their expertise and experience as managers, as they consider the once-dominant system of Neo-Liberalism. Neo-Liberalism is itself a theory or ideology of organisations on a global scale.

Rudd defines Neo-Liberalism. We will later revisit other definitions, but Rudd's words give us a useful starting point. He mentions three key qualities. It is 'extreme capitalism', a form of capitalism taken to a logical extreme. He is vague exactly what extremes he means here; but for him, capitalism per se is not discredited. The problem is a linear mentality, which pushes one tendency to its limit, without counter-balancing forces or measures.

He accuses it of being driven solely by 'greed'. Neo-Liberals prefer the terms 'profit' or 'self-interest'. But even if we avoid loaded words, the basis of this linear form of capitalism is sufficiently agreed. It trusts a single motive; self-interest directed to profit, to produce the best, most viable form of the world economy.

Rudd's first descriptor lies at the front and centre of this orthodoxy: 'free-market fundamentalism'. Neo-Liberals would baulk at 'fundamentalism'. It implies that its major thinkers are more like religious leaders, making irrational appeals to reason. But faith in the operations of the 'free market', whether based on reason or not, is core for Neo-Liberals. They believe that markets should be left to regulate themselves. Governments of all kinds at all levels should withdraw as much as possible.

A potential contradiction in this theory interests us as researchers of management and organisations. Neo-Liberalism is among other things a theory of management and organisations: how governments should manage the organisations within their sphere of influence. But the theory's main premise is that they should manage as little as possible.

If this principle were applied to corporations and other organisations, it would say that CEOs should try *not* to manage. They should trust in self-regulation below as well as above. They should be anarchists. But as is abundantly clear, they are nothing like this. The mirror image of Neo-Liberalism is called 'Managerialism', the doctrine that the meticulous practices of management and control should be applied to all forms and levels of organisation, irrespective of what they try to do.

CEOs invoke Neo-Liberalism to control governments above them as well as subordinates below them. In a mirror-image, Neo-Liberal Presidents like Mexico's Vicente Fox seek to attract and reassure multinational investment by talking like CEOs (Hodge and Coronado 2006). Either way, this creates a severe disconnect between different spheres of management, between macro-levels (national and global spheres) and micro-levels (companies and other organisations). Even before the crisis showed that this theory may not be the best way to run the world, it was already split by contradictions, between prescriptions and ideologies at different levels.

Rudd calls it a 'brand'. This sees it as being more like the object of a marketing campaign than a coherent theory or policy. We will not follow him all the way here. But we will not assume that there is a single, coherent theory here which may be shown to be right or wrong.

Contradictions are not fatal for a good marketing campaign. On the contrary, they are often of the essence. What matters is how they come together, how they are managed, and how ultimately they connect with the real world. Marketing campaigns can defer the moment of truth. They cannot evade it forever. Rudd sees the current series of crises as a moment of truth. On this point we agree.

He gives Neo-Liberalism only a short history, 30 years. British PM Margaret Thatcher came to power in 1975. Ronald Reagan became President of the USA in 1980. Rudd's estimate of the start of the neo-liberal revolution is defensible. He focuses on its current, sudden, and dramatic collapse, but he describes an equally sudden emergence. What appears that suddenly can collapse equally quickly.

He frames this history in terms of a 'crisis'. In fact, he talks of an escalating series of interlocking crises, in which each crisis triggers off another in an adjacent sphere of life. Boundaries that were thought to exist between these spheres were swept away. Differences between the larger and smaller scales did not survive. These are cascades of change that are so interconnected and rapid they are unstoppable, once the chain reaction begins.

It is not news to Neo-Liberals that the world is in a chaotic state. On the contrary, they claim that globalisation and its unpredictability constitute the natural environment for the definitive triumph of their theory, the One True Way to manage chaos. This crisis shows that Neo-Liberalism had not grasped the real principles of chaos as well as they had thought. It produced a catastrophe they could neither predict nor control.

But the idea that globalisation is now the inescapable condition for all businesses and governments has not been discredited by the crisis, on the contrary. All that has been discredited has been the idea that Neo-Liberalism had a good understanding of globalisation and chaos. The dominant theories of management that accompanied Neo-Liberalism, and were sustained by its supposed triumph, are also exposed to new challenges and criticisms.

Yet attacking these ideas is not enough, on its own. Now it is time for reconstruction, to develop new and better ideas about management and organisations. That is what we are trying to do in this book.

***

We take a particular angle on management and organisation studies: from below and from one side. From this angle, we can see fissures and contradictions of the dominant system with a clarity that is not so easy from above. From this place we notice many attitudes and values ignored by the gaze from above. These attitudes and values, the people and practices that embody them, the soil they grow out of, form the basis for a rich set of alternative ideas for rethinking organisation in today's world.

We use the term 'Larrikin Principle' to encompass a range of features that hang together in this alternative. We will not begin with full definitions of our key terms (Neo-Liberalism, Managerialism, Larrikin Principle). We prefer to allow definitions to emerge, to do justice to their richness and complexity. Yet, as we just did with Neo-Liberalism, we give a preliminary account as our starting point.

The term 'larrikin' is associated with Australia, and we use this context to help understand it, but the Larrikin Principle is not confined to Australia or English-speaking countries. Nineteenth century Australia, when the word first appeared, was nationalistic, xenophobic, racist and sexist. All these attributes coloured the conception, and congealed into a stereotype. Similar things happen to stereotypes in other countries. The 'typical Mexican', the 'typical Brazilian', the 'typical Yank' are potent ways of failing to understand the respective peoples and nations. Our version of Larrikins is coloured and inflected by the complex realities of today, post-modern, multicultural, gender-aware citizens of the world, a core part of our strategy to illuminate the issues of organisations throughout the Neo-Liberal world.

Paradoxically, many Australians reject the Larrikin Principle. Many non-Australians show more of it than do most Australians. None of its features is exclusive to larrikins or Australians. The principle came to Australia from a diverse global culture, and in this era, it reconnects with this scope and diversity. Instead of a symbol for a single national identity, for one small nation on the global scene, we use it to see existing and potential connections across this now profoundly connected globe.

Aussies still like to think it is typical of them. It connects with what they think of as their convict past, when their national character was shaped by opposition to the dominant, repressive British rule. This was not a typical post-colonial story. Larrikins did not rise up in arms and throw the coloniser out, as Mexicans and Americans (though not Brazilians) did. They developed a distinctive low-key strategy, beating and joining their oppressors.

In this sense Larrikins have a laid-back style. They are irreverent towards authority, bending or breaking rules if they do not see their point. They expose 'bullshit' wherever they find it. They adapt to new challenges with whatever comes to hand, with pared-down efficiency that gets the job done better than following the rules does. These qualities were born in frontier conditions the past, still needed in post-crisis organisations.

The Larrikin Principle is still alive and well in Australia today, in popular culture and in the world of business, even though a few decades of bipartisan Neo-Liberalism drove it underground. It acts through women and men at every level of every organisation, and their voices will weave throughout this book, reaching across the globe. Here are a few Australians to begin with, giving their take on the Larrikin Principle:

> Robert: It's irreverence or seemingly irreverence for people in authority. A readiness to break rules rather than accept them rigidly. That this is the way things should be done. The larrikin will look and say 'Oh yeah but there's a better way of doing it. We won't do it that way because we can do it better'.
>
> James: I think it's a more efficient way than, in some respects, the total bureaucratic way because ... which unfortunately I think is the way we're going. The level of rubbish and bureaucracy in our organisation's gone ballistic. If they could cut that out, we would become a much more efficient organisation and they wouldn't have to sack so many people to supposedly make us more efficient.
>
> Terry: One point I'd make is that if I'm in a situation where I can bend the rules to get a better result, one thing that would stop me doing that is if by bending that rule I felt like I was queue-jumping or disadvantaging someone else... Yeah, the larrikin thing has within it the concept of a fair go as well as anti-authority and not observing petty rules and that sort of stuff.
>
> Liana, a Brazilian-Australian: [Australians] accept more that they have to follow the rules and the normal bureaucratic process. On the other hand, it is interesting to observe that in Australia – I saw the statistics – for example, if in Brazil it takes a year to start a company, in Australia it takes a day. It is a country where bureaucracy has been eliminated. Just compare that in Brazil we come and say 'Ah, this document needs three stamps, etcetera.' Here it is much more simple. If someone signed it you believe.

In the book that follows we will expand and elaborate on these points. First comes the attitude to authority. The larrikin is only 'seemingly' against authority, as Robert says: against mindless respect for authority that is counterproductive. Second, the means, bending rules, relying on informal systems, is not opposed to rules as such, just to over-elaborate, rigid rules that are inefficient. Third, as Terry insists, these attributes are framed within a strong ethical framework, which rests on a respect for the rights of others, from the top to the bottom of organisations, inside as well as outside. Fourth, as non-Australian Liana observes, in comparison with her native Brazil, the key to easing the process is less defiance against rules than trust which does not need them.

We argue that these principles make for happier, more effective people in happier, more successful organisations, contributing to a better world. That is how we will use Larrikin Principles, as guidelines for inventing new and better forms of organisation, in a more just and less dangerous world.

***

The Larrikin Principle is a way of being critical, but we need to say more than that about it. Critics are not always welcomed, especially by the dominant. In times of peace and stability, criticism may seem unnecessary. In turbulent times it can seem dangerous. Either way, criticism can seem to interfere with the main business of managing. Critics are resented as 'troublemakers'.

Against this common perception, we argue that criticism is vital for any organisation, in its larrikin and other versions. We start out from the origins of the Greek word *krino* which means deciding between alternatives. From this word came two lines. 'Discrimination' in English carries one line, similar to 'criticism'. 'Discern' and 'certain' also belong to this family. 'Crisis' comes from the other line, referring to the objects of criticism. In medicine, this is a turning point in diseases that requires discriminating, critical eyes. Patients are 'critical', at a turning point between life and death, sickness and health. Doctors who are 'critical' see signs of health or disease that others may miss.

This history of words carries lessons about criticism and why and when it is necessary. There is a deep connection between criticism and crisis, as the history of these words suggests. Times of crisis unleash a range of criticisms. Signs and assumptions previously taken for granted are scrutinised. This can be portrayed as destructive, but it is needed to restore health. Crisis generates the need for criticism, which becomes feed-back, to help understand and manage the crisis. A system or organisation without strong critical loops is dangerously unprepared for any change in business as usual.

A major crisis, such as the present case, needs flourishing, diverse criticism, and alternative ideas about what went wrong and what else might be tried. That is precisely what is happening now across the globe. Business life even in normal times constantly negotiates turning points, requiring small or large adjustments by managers capable of discrimination and judgement, who are 'critical' in our sense.

In the field of Management Studies the Larrikin Principle would be placed into a stream called 'Critical Management Studies'. This group gained this label only recently. A manifesto by Alvesson and Willmot (1992) called it a loose grouping in Business Studies, created by a flow of left-leaning academics from outside management, from sociology and related disciplines. We share many of the qualities of this group. We agree that organisations and their contexts can only be understood as social forms. Disciplines like sociology, anthropology, history and semiotics are part of the disciplinary mix of Critical Management Studies.

Using Rudd's brief history as a framework we note that Critical Management as a named movement arose almost exactly half-way through Neo-Liberalism's 30-year lifecycle, though some management academics had been critical for many years previously without the name. Co-incidence?

Since its inauguration, this stream has grown in number and standing. It remains a minority position in Management, yet it now has a place. No Neo-Liberal luminaries have said in public that the orthodoxy felt an increasing need for this counter-balance, but that is what has happened. The advent of the crisis does not mean that there is no longer a role for critical forms of management, on the contrary.

Yet, even within Critical Management, there is a sense that the movement is not getting its message across as it should. Martin Parker, a British Critical Management author who writes like a larrikin, takes them to task for their ponderous style:

> The arcane nature of many of [their] arguments, the endless debates between neo-Marxists and Foucauldians, realists and post-structuralists, and a typically academic emphasis on the importance of Big Theory means that most of this writing is rarely read outside the academy (2002:14).

This 'arcane' style has its reasons. It flows from attitudes of respect for decorum which do not paralyse larrikins like Martin Parker. Or us.

There are significant differences within Critical Management. These do not weaken it but give it more of the diversity it needs for these times. For instance, Stewart Clegg, an influential figure in Critical Management Studies, offered an insightful criticism of the movement as it appeared to him in 2008. 'A spectre haunts this collectivity, and it is the apparent dearth of any alternatives to capitalisms' – note the emphasis on the plural (2008b)

Clegg argues here against simplistic negativity, not against a critical gaze. He wants a greater recognition of the complexity of capitalism on the agenda. Without this, the critical function will miss its primary purpose – to see differences. This is the kind of 'criticism' that the world needs from management, 'critical' or otherwise.

We begin with a simplified diagram of the relationships we see between Critical Management and the Larrikin Principle, and between both of these and Mainstream Management. We then show two ways in which this edifice of knowledge might be related to the world of organisations, which it supposedly exists to explain.

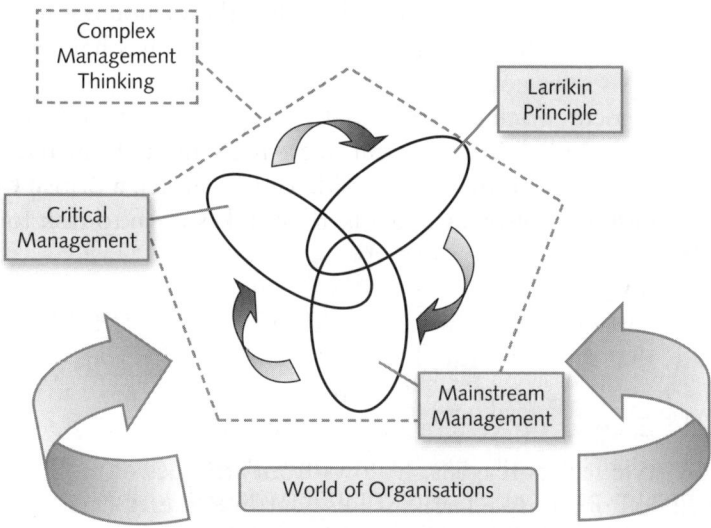

Management Thinking in a Complex World

This diagram is built around one idea of Management Studies; that it is a single field, which feeds ideas into the world of organisations, from which it receives data and problems. However, we set this simple story in a more complex dynamic picture of management studies and relationships with the world. In this richer picture, the three forms of management thinking jointly form a many-stranded system of knowledge (which includes more than the three we have included). Each shares some things, but the differences are equally valuable. They make the system more dynamic, more adequate to the diversity of what they attempt to explain.

As well as shared points, we also show linking arrows. Debates and dialogues flow across the boundaries, slowly changing each of them, as people in each field become more conscious of ideas and issues they may be taking on board from the others.

Critical Management has been in this dialogue for at least 15 years; long enough to become a dynamic mix of Management and Criticism, to different degrees in different writers. Mainstream Management seems less affected by the dialogue, but this may only seem so because Mainstream Management has greater inertia. The diagram implies our sense that Management has already been subtly altered, and will be more so.

The Larrikin Principle is a newcomer to the system, closer to Critical than Mainstream Management. Yet it approaches Mainstream Management from a different angle, making different connections. It is interested in learning from Management and Organisation Studies, in all forms. It brings distinct conceptual and analytical tools, which both Critical and Mainstream Management might find useful.

***

This diagram obscures one key issue that divides the different critical traditions between two poles. At one extreme, positions should be kept as separate as possible. Any trace of compromise or complicity with the opposite tendency is rejected. One slogan for this approach was the powerful phrase of Afro-American radical feminist Audre Lorde: 'The Master's tools can never dismantle the Master's house' (1984:110).

We have a mixed response to Lorde's famous phrase. On the one hand, we recognise that there are situations so extreme that this stance is justified, and the Larrikin Principle seems inappropriate. To take just the case of international aid: critics of Western aid, from Schumacher (1973) to Easterley (2006), agree that Big Plans and Big Technology from the West do systemic damage to recipient societies. Western aid, packaged with Western ways of thinking, proves to be a poisoned chalice.

But the combination is not some metaphysical condition that cannot be resisted. On the contrary, the chains that bind the package together can be named, and resisted. They are Neo-Liberal policies and Managerialist methods.

As another instance, ethnography as a research method has a long history as the Master's tool. In the days of European Empire, ethnographers went out amongst the 'natives' to report back to the metropolis on what made them tick, to make them more manageable (Hodge 2008). Palestinian-US writer Edward Said called this practice 'Orientalism' (1978). Yet in Said's account, as in ours, the worst trick of Orientalism was to claim that this knowledge, which they appropriated from 'native' experts, was always only theirs. The problem was not ethnography as such, but ethnography as theft.

Lorde is herself in some ways a kind of larrikin. She called herself an 'outsider', a radical Black feminist and lesbian, a critic of dominant critics of her time. Larrikins have affinities with outsiders. We see her as our kin. Yet our larrikin attitude to the idea of 'the Master's tools' is completely the opposite. Larrikins do not assume that the Master really owns the tools he claims to. They doubt that the Master knows how best to use them. They are happy to give them a go.

After the crisis, as before, larrikins doubt whether Neo-Liberalism ever understood globalisation, in spite of its claims to be the One True Way to manage it. Globalisation remains open to larrikin hands and larrikin theory, too vast and important to be left with the former Master. Managerialism claimed to know the best way to manage organisations, small and large. Larrikins doubt that, too. The dominant system offered itself as a single system, for good or ill, like it or not. The Larrikin Principle is a way of probing such claims.

***

So far, we have emphasised contradictions in the dominant system, perhaps giving the impression that its contradictions will be enough to destroy it, or that the Larrikin Principle has no contradictions. On the contrary, Marx identified huge contradictions in capitalism in the middle of the 19th century, but these have not destroyed it yet. Even the current crisis is likely to pass with capitalism changed in various ways but still intact. Rudd and Obama still expect to govern capitalist nations in a capitalist world. Far from the Larrikin Principle avoiding contradictions, it needs to incorporate them, to understand and cope with the role of contradiction itself.

To describe a world order containing both power and contradiction, we adapt the concept of the Ideological Complex, developed by Hodge and Kress in 1988, (ten years after the birth of Neo-Liberalism in Rudd's history):

> A functionally related set of contradictory versions of the world, coercively imposed by one social group on another on behalf of its distinctive interests, or subversively offered by another social group in attempts at resistance in its own interests. (1988:3)

In these terms, contradictions in Neo-Liberalism/Managerialism are not necessarily a sign of weakness or imminent collapse. On the contrary, they offer its ideologues countless opportunities to defend their system from different angles, saying that black is black or white depending on which is most convenient. Given this inherent slipperiness in the dominant ideology, the Larrikin Principle must be equally mobile, drawing on its own contradictions, if it is to perform its core task of exposing 'bullshit' in all its forms.

***

Our approach to issues of organisation and the crisis of Neo-Liberalism is distinctive in its use of theories of chaos and complexity. We do not claim that the Larrikin Principle absolutely requires such a framework. The Larrikin Principle comes from below. At present, theories of chaos and complexity come from above and outside of management itself. Yet, we believe there is a deep affinity between these theories and the Larrikin Principle. We also see a larrikin approach to these ideas as the best way to open them up as complex, effective tools for thinking about organisations, from below and above.

Chaos theory and complexity science are more or less respectable today because some very good scientists and mathematicians have taken them up. But this respectability has come at the usual price. Academic proponents of the theories have packaged them as science, dismissing everyday talk about chaos and complexity, without the mathematics, as vulgarisation and misunderstanding.

As larrikins we are suspicious of appeals to authority used to exclude ordinary people. Ideas of 'chaos' and 'complexity' had a long, rich history before these theorists took them up. We welcome contributions from these scientists. But we believe that only through larrikin willingness to try these ideas out will we be able to understand them in practice, following them into the life of organisations.

In fact, management and organisation studies have taken these ideas up more strongly than have most other social sciences. Business guru Tom Peters wrote a best-seller in praise of chaos (1987). Billionaire Bill Gates promoted chaos theory as part of what he saw as a revolution in conditions of business (1999). As larrikins we do not criticise them as popularisers. We wish that more followed their ideas. So far, they feed rhetoric more than thought, but that can surely change.

Here we sketch a small took-kit of ideas regarding chaos and complexity, which we will develop further as the book progresses.

1. *Far-from-equilibrium dynamics*. This is more a framework than an idea. It has been developed most productively by Ilya Prigogine. (Prigogine and Stengers 1984). As a scientist, in a muted way, Prigogine had some larrikin qualities. He was a nomad, an intellectual boundary-rider. He escaped the Russian revolution with his family for Belgium, from where he won a Nobel Prize. Later he moved between Europe and the USA, a mobile and productive 'larrikin' scientist.

In Prigogine's theory, things behave differently under different conditions. Close to equilibrium, everything is fixed or moves slowly. Things are easy to control and follow linear logic. That is the world assumed by Managerialism, in which every action by managers, if performed correctly, will produce precisely the desired effect.

Scientists Davies and Gribbin offer a simple account of linearity:

> In physics, a linear system is, simply speaking, one in which the whole is equal to the sum of its parts (no more and no less) and in which the sum of a collection of causes produces a corresponding sum of effects (1991:38)

But further away from equilibrium, things change dramatically. Instead of linear models being the best and only form for scientists and managers to use, non-linearity becomes pervasive. Predictions become difficult or impossible. Causes can act over great distances, or not at all. They can produce larger or even opposite effects.

One famous instance of this is the so-called 'butterfly effect' (Lorenz 1993), captured in the image of a butterfly flapping its wings in the Andes, precipitating a hurricane in Montana. Minute initial differences produce hugely different outcomes. Crises and contradictions abound in a far-from-equilibrium world. This is a control-freak's worst nightmare. The global financial crisis is one such case.

But Prigogine has good news for organisations. The wayward world at the edge of chaos produces all its most interesting and functional forms. He calls them 'dissipative structures'; open and dynamic systems which feed off the energies of chaos. Some of these complex systems are remarkably stable. Chaos underpins order, he claims: not always, but often enough to have produced life on earth, and all the achievements of humans and other biological forms. Organisations at equilibrium are easy to control but inert. All successful organisations today find order out of chaos. Globalisation is far-from-equilibrium dynamics at work.

Adam Smith, patron saint of Neo-Liberalism, had already intuited something along these lines in the 18th century, in his celebrated theory of the market as an 'invisible hand' (2008/1776). Prigogine's idea of dissipative structures provides support for the idea that elaborate structures of regulation and control may be counter-productive in far-from-equilibrium situations. Yet, the same idea exposes the contradiction in the dominant theory, between Neo-Liberalism's rhetoric of chaos and Managerialism's obsession with control.

In these conditions, the larrikin phrase 'have a go' comes into its own. It is aware of difficulty and complexity, able to cope with uncertainty, yet committed to action.

2. *Complex adaptive systems.* This idea nests comfortably within a larger Prigogean framework, but has flourished in its own right, to the extent that some authors prefer to call the whole field 'complexity science'. Perhaps it has the advantage of being less threatening. In this book we do not define them strictly, or emphasise the differences.

An influential proponent of complex adaptive systems is yet another Nobel Prize winner, US physicist Murray Gell-Mann, who describes a common pattern found in biology, ecology, psychology, learning, thinking, finance and computing:

> A complex adaptive system acquires information about its environment and its own interaction with that environment, identifying regularities in that information, condensing those regularities into a kind of 'schema' or model, and acting in the real world on the basis of that schema. In each case, there are various competing schemata, and the results of the action in the real world feed back to influence the competition among those schemata. (1994:17)

Gell-Mann sees such systems as generative: 'complex adaptive systems, it turns out, have a general tendency to generate other systems' (1994:19). Complex adaptive systems are included in what Prigogine called 'dissipative systems'. They are such a vast, unknown field of study that both theories combined still only provide a rudimentary handle on what is involved. This is a surface that we can only hope to surf over in this book.

3. *Cybernetics* is one of the most powerful ideas of the 20th century, foreshadowing both chaos theory and complexity science. Gell-Mann's account of complex adaptive systems describes a classic cybernetic system. In his autobiography, Gell-Mann describes how he met the inventor of cybernetics:

> 'a great but eccentric mathematics professor at the Massachusetts Institute of Technology, Norbert Wiener, who as a child had been considered a prodigy, and never got over the need to show off in bizarre ways' (1994:72).

Behind this acerbic comment, Wiener emerges as an outsider with some larrikin qualities. His genius was to see a simple idea crossing biological, social and engineering systems, captured in the Greek *kybernetes*, a helmsman. The word already covered political and social systems. The English verb *govern* comes from the same root. Wiener applied the same principle to all forms of organisation in order to explain the extraordinary complexity of all human skills.

A key idea in cybernetics is 'feedback', the idea that information is fed back into the control system to influence behaviour. Wiener (1948) distinguished between two kinds of feedback. 'Negative feedback' cancelled a positive tendency, whereas 'positive feedback' increased it. This is confusing because 'negative feedback' often entails what seems to be the positive outcome of stabilizing the system, whereas 'positive feedback' produces runaway effects which escalate into chaos. To avoid this confusion, we refer to feedback as 'dampening' or 'accelerating'.

Wiener explains one kind of chaos and how to manage it. A form of butterfly effect is produced by accelerating feedback chains. The cascades of crises Rudd pointed out are accelerating feedback loops running through a system with devastating effect. Yet, without a mix of accelerating and dampening feedbacks, control systems would not work. To take a *kybernetes* steering a boat: small changes in the position of the helm are magnified into large effects on the ship's direction via the rudder by means of a chain of accelerating feedback, to counteract (damp down) its swings of direction.

Management and organisation studies claim cybernetics as their own tool, so we take larrikin pleasure in liberating it. We use it to dismantle the too-static edifices of Managerialism and Neo-Liberalism. For instance, 'feedback' is a common word in management, backing up managerial systems of control. Managers claim to 'consult' prior to introducing changes, often just to tick the 'consultation' box. This tactic is meant to make the workforce feel that their views are respected, to quieten opposition (dampening feedback). In this form, it may increase alienation (accelerating feedback).

As Wiener recognised, cybernetics is a non-linear science of complexity, not a set of quick fixes. 'Cyberspace' is the vast, chaotic field of cybernetic processes. It is not only 'out there', in the Internet. It is also inside countless processes; from car engines to banking systems, in all social and economic systems.

4. *Fuzzy logic.* Another larrikin/chaos thinker is Lotfi Zadeh, born in Iran and now a US citizen; another nomad thinker who specialised in crossing boundaries. Zadeh was an engineer, an expert in the cybernetics of control systems. However, he found that the crisp logic prized by engineers, scientists and managers broke down in the face of highly complex systems and conditions, including all systems involving humans.

His solution was fuzzy logic in which boundaries around concepts stay fuzzy, and statements may be only part-true. The more we push for absolute precision, he argued, the more meaningless or irrelevant our schemes become. He captured the core hypothesis in what he termed the Principle of Incompatibility:

> Stated informally, the essence of this principle is that as the complexity of a system increases, our ability to make precise and yet significant statements about its behaviour diminishes until a threshold is reached beyond which precision and significance (or relevance) become almost mutually exclusive characteristics. (1973:28)

Fuzzy logic is fundamental to larrikin wisdom. It is contained in the common Aussie phrase that alarms all uptight managers: 'She'll be right'.

5. *Fractals.* Belgian-US mathematician Benoit Mandelbrot's theory of fractals (1993) has been a poster-child for chaos theory, with its computer-generated images being reproduced on everything from T-shirts to book covers. As one instance, the cover of a recent textbook *Understanding cross-cultural management* (Browaeys and Price 2008) has an attractive image inspired by fractals. At the centre, a shape formed by four circles of dots expands exponentially to reveal that each component dot is really an image of planet earth.

This image illustrates many things about fractals. They describe similar patterns across different scales. This is a useful function for analysis. They guide heuristic (discovery) questions. For instance, one key question drives our analysis: what is Neo-Liberalism like at global and local scales? Initially we were struck by difference. Managerialism seems to have the opposite attitudes to control systems compared to Neo-Liberalism. On closer inspection we find a complex pattern of contradictions, between rhetoric and practice, linear power and claims of non-linear freedom. A fractal, Mandelbrot insists, is not a simple, mechanical pattern replicated at every level. Complex, irregular patterns are reconfigured at every level in fractal series. What is replicated is not a simple ideology, but an Ideological Complex.

This image seems to have come from the marketing department, not from the authors. There is no mention of fractals in the book, not even to acknowledge the source of the image. In this case, the publisher picked up fractal theory as if it were purely decoration, with no intellectual content. As larrikins, we take fractals back from marketing, and use them.

6. *Power laws*. One old idea that chaos theory has taken up is the concept of Power Laws. At its simplest, this is just the idea that one or both axes of a normal two-axis graph may be an exponential number, giving rise to a characteristic curve that begins or ends flat, but increases ever more steeply. A Power Law refers to phenomena that produce this shape.

Power laws have become more interesting because thinkers like Mandelbrot and Gell-Mann have been interested in them. Fractals follow a Power Law. So do many other phenomena. George Zipf (1949) applied it to sets of words in literary works, and the seemingly unrelated theme of the size of cities. Within a given nation (Australia, Brazil and Mexico for instance) city sizes all follow this curve, in spite of their different histories. Danish physicist Per Bak (1996) sees it as a signature of systems that can become critical, and generate chaos or complexity.

7. *Three body systems*. For some writers, chaos theory was born of an analysis by 19th-Century French mathematical genius Henri Poincaré of the so-called 'three-body problem'. The problem was seen as a big deal because the three bodies it involved were the earth, the sun and the moon in our solar system. According to Newton's definitive scientific breakthrough, these must be strictly governed by Newtonian mathematics.

Poincaré rained on the Newtonian parade. He showed that if there were only two bodies, say the two big ones, the sun and the earth, we could predict precisely where each would be after countless revolutions. But if we add just one more body, say the humble moon, then the mathematics breaks down. The further into the future we look, the less we are able to predict. He used the Master's tools (Newtonian mathematics) to dismantle the edifice of the Newtonian system.

That means that, had we stayed with two bodies, our predictions would have seemed certain, yet would have become increasingly wrong, and we would not have suspected it. Better to be less certain, and more right. It is Zadeh's paradox again. Fuzzy logic is more precise than excessive precision taken too far.

This way of thinking incorporates other traditions; it does not eliminate or replace them. Poincaré did not attack the number two, or all numbers up to infinity. Two is a powerful, productive number. Two heads are better than one and, as Bob Dylan noted, one hand clapping makes no sound. Problems come from the exclusive, reductive use of twos in unreflexive binary analysis.

One lesson of three-body analysis is that any pair is always part of a larger set. However small and insignificant the third body is, like the moon, we should not ignore it. It always has effects, which become ever greater over time. We used a three-body analysis, but without the name, in the diagram 'Management thinking in a complex world'. The Larrikin Principle was the small third body. It opened up an infinite, unpredictable set of possible relations between mainstream and critical management.

The talismanic larrikin phrase 'Fair go, mate' carries a version of 3-body analysis. It is what larrikins say in the face of injustice, where the many or the powerful oppress the small and the weak. In such exchanges the larrikin is like the moon, the third body, outside of the battle between the two in conflict, yet able to change its outcomes, over a longer period, in surprising and unpredictable ways.

* * *

We authors are all academics, doing the work that is typical of academics (reading, writing, researching) in the habitats typical of our species. Most books on organisations and management are written with a different default idea about work and organisation. Yet, we wanted to overcome this sense of difference and disconnection. Personal experience is the best teacher of complex lessons. We needed to be able to draw on our own.

Our team consisted of four people from four different nations, with four different backgrounds; two men and two women arbitrarily dumped in the same workplace, in Australia. We were products of globalisation, a hybrid, transnational team, facing complex, demanding tasks. Different personalities, abilities and aspirations had to be balanced and integrated: a typical organisation problem after all.

We used ideas from chaos and complexity to mediate between personal and generic. We framed our project in three-body terms, simplifying the full many-body situation of the world while still capturing some of the irreducible complexity of global relationships and flows. Our three bodies were Australia, Mexico and Brazil, chosen partly because they were important to one or more of us and partly because they occupied significant positions on the peripheries of the global system. We understood them all as revolving around the sun at the centre of the current global system, the USA.

Fractal analysis and fuzzy logic complemented this three-body analysis. Our team, a Mexican, and Brazilian, a Canadian and an Australian, all living in Australia, can be seen as a fractal of the international relationship. As a heuristic device, fractal analysis invites us to look for equivalent complexity on smaller scales, while fuzzy logic allows us to describe it more easily. If we look, we find. If we can describe it easily, we can think with it better.

Gabriela and Fernanda, for instance, are both Australian and Latin American, in different ways. Both were born in a Latin American country and lived there for many years. Both migrated to Australia, married Australians, and became Australian citizens with dual nationality. In terms of the crisp logic of national stereotypes, they are still just one thing, Mexican or Brazilian. Both should be like all other Mexicans or Brazilians, and completely unlike all Australians.

To judge them like this is rational, according to the dominant (linear) definition of nationality. But this way of thinking misses the complex reality of what they are and may become. It creates problems when it comes to thinking about their possible roles in a cross-cultural team whose cross-cultural depth is an important resource.

Gender needs to be treated similarly. In sociological categories, Gabriela and Fernanda are both female, as Greg and Bob are both male. In linear thinking, there is nothing more to be said. But if the fractal of the team combines male and female, each participant can also be seen as combining, in some way or to some degree, both male and female.

On closer inspection, most of the activities of men and women are not exclusively gendered: neither male nor female. In fuzzy logic, the women are female and not-female, the men male and not-male: fuzzy males, fuzzy females. Without the idea of the fuzzy gender of fuzzy larrikins, we would have found it hard to use the Larrikin Principle as we wanted to. The larrikin in Australian culture was a male stereotype, seemingly restricted to a certain class of man, not a helpful way to re-think relationships in organisations today.

To get round this problem we coined a new term, *larrikina*, the feminine form in Portuguese and Spanish, the two main languages of Latin America. Larrikin(a)s are cross-gender as well as cross-cultural. There are still cross-gender differences, but language should not be a barrier to seeing and talking about all the ways that experience is and is not gendered.

***

This project, in the form it took, needed more than larrikin enthusiasm. The team was fortunate to win a substantial grant from the Australian Research Council, to fund the data collection for an ambitious project. We were grateful for this support, which we used to collect stories about organisations from ordinary people in Australia, Mexico and Brazil, including transnational Mexicans and Brazilians living in Australia. We analysed them to build up a richer story.

We finished with over 2,000 pages of transcripts, over 700 stories and 1200 reflections from over 100 people from our three parts of the global world. We scanned the media for stories from all three nations. We added a comprehensive set of management textbooks to our database. We were helped in dealing with this mass of materials by two wonderful Research Assistants, Sandra Klinke, a Brazilian with administrative experience, and Laura Calderón de la Barca, a Mexican cultural analyst and psychotherapist, who both also discovered their inner larrikina. In the final stages we were joined by Beatriz Cardona, a Colombian with her own complex history and contribution.

The fractal principle opened up a wealth of illuminating data, at many different levels. We could see apparently minor incidents recounted by countless ordinary people, in Brazil, Mexico and Australia, as carriers of complex and important meanings about processes in organisations and nations in a global framework. Personal stories related by members of our team provided further rich data. So did team film evenings.

We held our team meetings once a month, booking a meeting room in U9, a building which housed a group of academics in the School of Management. But we soon learned that we had to close the door properly. 'There's too much laughter here to be a research meeting', laughed Anneke. Louise, whose room was closest to our meeting room, just closed her door when our meetings started, without complaint but with a good line in understated satire. We tried to be more constrained and considerate to others, but we did not fully succeed. Besides, laughter is a key marker of larrikinism.

These colleagues were not only remarkably tolerant. Over time, they came to shape the themes of the book, and inspire its conclusions. As colleagues they were warm, supportive, and constructive. As we also came to learn, they were the ones whose efforts, dedication and humanity often carried a School that was staggering under the weight of Neo-Liberal dictates and managerialist solutions. Most came from management positions, being far more competent administrators than most academics (and many of their current bosses). They showed us the complex, low-key reality of the Larrikin Principle at work in organisations. All we had to do was build the theory around their example.

***

We end this preface where we started, with the words of Australian Prime Minister Kevin Rudd on the implications of the crisis:

> The time has come, off the back of the current crisis, to proclaim that the great neo-liberal experiment of the past 30 years has failed, that the emperor has no clothes. Neo-liberalism, and the free-market fundamentalism it has produced, has been revealed as little more than personal greed dressed up as an economic philosophy. And, ironically, it now falls to social democracy to prevent liberal capitalism from cannibalising itself. (2009:25)

Here Rudd targets not just an economic doctrine, but also the myth that has sustained it. In this myth, Neo-Liberalism is the hero, riding the white horse of the Free Market, with the sword of Managerialism in his hand, and his allies the Aspiring Poor, defeating the dragons of Inefficiency and Corruption. The King, Globalisation, promises him his daughter, Unlimited Profit, as a reward, plus his kingdom, the planet itself. Australia, Mexico and Brazil are lost in the crowd cheering his victory.

We responded to this myth sceptically, as larrikin myth-busters. Our stories from below and outside the centre painted a different picture. In our stories, Neo-Liberalism is more likely to be a villain or an obstacle than a hero. He needs the heroes of our stories if he is to cope, along with dirty tricks not mentioned in his myth.

We also offer a counter-myth. Its hero is the larrikina, cross-cultural, cross-gendered, postmodern saviour of organisations everywhere, carrier of ambiguity, anomaly, disorder, with global roots in deep time. His horse is Cross-cultural Alliances, her sword Informal Practices, his allies Social Capital, Culture and Networks, her enemies Linearity and Injustice. His world is complex, dynamic and unpredictable, and she does not try to own or control it. Australia, Mexico and Brazil are three of many friends who meet along the way, a growing crowd in a journey that is only just beginning.

Chapter 1

What's wrong with business education?

We have many reasons to examine business education in terms of the Larrikin Principle. It is a crucial site in the cybernetic cycles whereby practices and ideologies of management are formulated and have effects. The next generation of managers are being taught in business schools, which frame what they are being told about the world of business and its pasts and futures. In cybernetic terms, this becomes a key site for seeing what versions of the world particularly need to be examined and queried, if things are to be different in the future.

The fact that these activities take place in universities adds to their interest as far as we are concerned. Universities and university systems are major forms of organisation in today's societies. They are the major organisation, perhaps the only one, which teachers, students and business people have all experienced. They could be a common reference point, to ground discussions and debates for current and future managers and their teachers. Yet they are hardly mentioned on management courses.

Perhaps this rich resource is deliberately left to one side out of a sense of propriety. Some may feel that it would not do for teachers and students to talk about the institution that surrounds them all. Perhaps current managers feel that they have now gone beyond that stage in their lives. Maybe they do not want to remember when they too were students in an imperfect institution, or to speak ill of their Alma Mater.

Whatever their reason, the Larrikin Principle lacks the sense of decorum required to observe this boundary. We see it as vital to our critical task to focus directly on the relations between universities and business life, to continually stray across this boundary.

1. The crisis in business education

Even before the financial crisis struck with full force, a sense of malaise gripped schools of Business and Management in universities across the globe. Leaders in Business Education were aware of it. There were many warning voices from critics of all kinds. It is just that they were not listened to.

For instance Stewart Clegg entitled a major address given in Brazil: 'And there are no truths outside the gates of Eden' (2008b). He applied Bob

Dylan's words from the 1960s to Business Studies today. They lack a moral compass and a coherent intellectual foundation, he said. Business *Schools* in institutions of Higher Education are becoming *Business* Schools that are subordinated to the current needs and interests of business. They have no thoughts on Eden, no vision of better ways of doing business to help create a better world. Instead of preparing students for a profession, they only equip them with techniques, to use as their future bosses require. (2008b)

In 2008, Philip Delves Broughton reflected on his experience only 4 years before (2008) studying for his MBA at Harvard Business School. He says that he was sensing even then a seismic shift. It pivoted around issues of ethics in business:

> In 2003, Harvard introduced a class called "Leadership and Corporate Accountability" to allow students to discuss the perils of chasing dollars down ethical sewers. (2008:4)

In one class, he reports a flash-point in a discussion of an argument that once would have encapsulated the ethical position of the classic capitalism of the 1960s:

> Many successful business people lived by one set of ethical issues in their private lives and a quite different set in their professional lives… Knowing that you could win the game of business playing all manner of tricks which you would never inflict on your spouse, children or friends made for a calmer, less complicated life.

This strategy, of solving ethical dilemmas through a total split between private and professional values, now seemed problematic to the class. In the 1960s this was seen as a way of removing complexity. Now, in an environment perceived as already highly complex, it would only make life more complicated.

Heavy criticisms also come from Management mainstream. As early as 1994, Henry Mintzberg, a distinguished Canadian management theorist, had denounced the dominant style of MBA, as exemplified by the Harvard model. Instead he argued that:

> much of (its) success is delusory, that our approach to educating leaders is undermining our leadership, with dire economic and social consequences. (1994:5)

Outline of a critique

Two other heavyweights, Warren Bennis and James O'Toole, published an influential critique in the *Harvard Business Review* in 2005 entitled: 'How Business Schools lost their way'. These writers begin bluntly: 'Business

Schools are on the wrong track' (2005:96). They are graduating students 'ill-equipped to wrangle with complex, unquantifiable issues.' This is the 'stuff of management', they say, out there in the real world. Yet back in Schools of Business, even the best, there is an obsessive but irrelevant focus on 'scientific' models. 'When applied to business – where judgements are made with messy and incomplete data – statistics and methodological wizardry can blind rather than illuminate'. (2005:99) The problem, they say, 'is not that business schools have embraced scientific rigour, but that they have forsaken other forms of knowledge' (2005:102).

Their critique contains many points that we develop throughout our book:

1. Complexity is the irreducible condition of the world of business decisions, in businesses themselves and even more so in the world that business has to deal with.
2. This complexity is ignored in dominant forms of business education and theory, making many real-life problems difficult to address or deal with.
3. A major source of this inadequacy is the exclusive use of linear, instrumental rationality, reinforced by the prestige of science, typified by the overzealous and endless promotion of 'models' in business texts.
4. The dominant linearity loses a variety of approaches and disciplines.
5. Critical thinking is vital within as well as outside the mainstream.

Bennis and O'Toole identify a number of disconnects which together show why they are so concerned, as everyone should be who is concerned with the future of business studies. A simple cybernetic diagram is a helpful way of clarifying some of the wider implications of their position.

A Cybernetic Model of Business Education

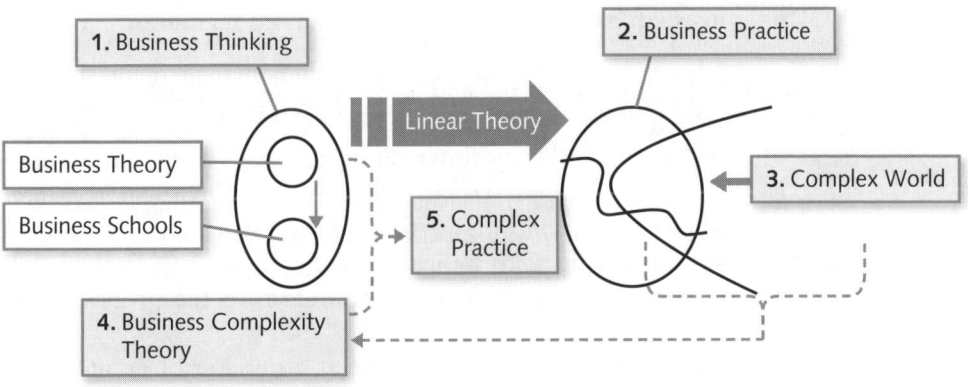

In *1, Business thinking*, we see Business Schools dominated by one brand of theory, a linear mono-discipline isolated from all other disciplines in the academy. No feedback loop links business practice and business education, thus it is out of touch with both.

There is one major absence from this picture, Neo-Liberalism. Clegg's critique named it as the model whose form of linear thinking turns all values and processes into cash terms, what he terms 'financialisation'. He blamed the over-balance of one discipline, economics, in the mix that has traditionally underpinned business studies (the others being sociology and psychology).

Distinguished Indian management theorist and educator Sumatra Ghoshal (2005) denounced these economic theories as 'bad theory': 'ideologically inspired, amoral theories with which business schools have actively freed their students from any sense of moral responsibility.' Ghoshal connects this with the recent scandals hitting the world of business in the USA, e.g. the collapse of Enron. He could equally apply it to the 2008 economic meltdown. In both, amorality underpins not success but catastrophe.

In *2, Business Practice*, we see the complex ways Business Thinking of this kind interacts with the world of business. Managers know they cannot afford to rely only on the edicts of linear theories, but they still operate in an environment dominated by Managerialism and Neo-Liberalism, learnt so well in Business Schools. The result is a split in managerial minds, dividing theory and practice, which they find repeated and exaggerated in the minds of the graduates they then recruit. This is 'bad theory', in Ghoshal's words, which creates tensions within the Ideological Complex.

In *3, Complex World*, we represent the all-pervasive intrusion of the complexity and chaos of the external world into business decisions and activities. The different facets of reality that make up this mix are studied by all disciplines in the academy, so that multi-disciplinarity is essential, as Bennis and O'Toole insisted. This multi-disciplinarity has to include science in all its forms.

Between 1 and 2, and between 3 and 4, we show no feedback loops which would allow the whole system to learn and grow. This is one price of linear systems of control with one-way flows. However, we inserted 4, *Business complexity theory* that grapples with the hyper-complexity of the interlocking systems of business and the world. We represent these theories and feedback links in dotted lines, to indicate that Bennis and O'Toole do not specifically recommend them. They are our solution to the problems they identified.

The Larrikin Principle can play a vital role at a number of points in this scheme. Larrikin irreverence allows the feedback that the system so badly needs. It offers the critical perspective that Bennis and O'Toole also exemplify. The Larrikin Principle is allied with chaos perspectives, yet in an easy, relaxed and unthreatening form.

The Management Education machine

Business education is a conveniently compact object of study, compared to other fields of education. It is especially easy to study because of the role of textbooks in pedagogical practice. Management textbooks in Australia play a more prominent role in business education than in any other field of humanities or social sciences. This is a well-organised global industry, ordered along good business lines by a small body of educational publishers. It is a formidable, well-oiled machine. The crisis has had no time to impact on it, but we do not expect it to interrupt its progress.

To give a sense of how this machine impinges on business education, using Australia as reference point, we look at a corpus of 12 textbooks which between them have penetrated most Australian schools of business and management.

Author	Publisher	Year	Pages	Editions	With aids
Brookes	Pearson	2004	461	1	Yes
Browaes	Pearson	2008	363	1	Yes
Davidson	John Wiley	2003	824	2 + 5	Yes
Deresky a	Pearson	2006	508	5	Yes
Deresky b	Pearson				Yes
Hill a	McGraw-Hill	2008	509	1	Yes
Hill b	McGraw-Hill	2004	582	3	Yes
Hitt	Pearson	2007	677	1	Yes
Hubbard	Pearson	2004	470	2	Yes
Robbins	Pearson	2006	820	4	Yes
Ryan	Pearson	2003	323	2	Yes
Waddell	McGraw-Hill	2007	542	1	Yes
Totals					**100% Yes**
Averages			507	2.1	**100% Yes**

Our sample of 12 textbooks has a number of generic qualities. First, they are marketed and sold as commodities. All have glossy coloured covers, and glossy paper. They are all physically substantial. We listed the numbers of pages, averaging 507, but this underestimates the bulk of these books. The average page size is 500 cm^2, double the size of an ordinary paperback (240 cm^2) If these text books had the same page-size as paperbacks, they would average a whopping 1,000 pages, for just one course unit.

Our point in making these calculations is to bring out a paradox in the nature of the material commodity. If the aim of this industry were to deliver the maximum information at the lowest cost, none of these text-books would be efficient. They are big in order to be expensive. They are so expensive that only one can be set for a course.

The pricing and marketing strategy is designed to remove choice within the course. Students at our university take four units per semester, roughly 10 hours per unit, which has to include 3 hours of classes. They also need time for assessment tasks, on average 3 hours per week. That leaves 4 hours per week over 13 weeks, 78 hours. If conscientious students read 20 pages per week closely, equivalent to 40 pages of paperback, they would still only be able to read roughly 260 pages in total, about half the average size of one text.

So, half of the text is meant to be redundant. Students are required to buy redundant material, and given no time, space or direction to read this or anything else. The machine's strategy is to offer more choice that can be taken up within its scope, yet cut off other sources of choice and diversity.

Nor is this choice theirs. The textbook is prescribed by course co-ordinators, not voted on by students. It is true that if a text does not work with one cohort, their experience may affect the course-co-ordinator's choice for the next year, but the decision lies with lecturers, not students.

Reflecting this reality, publishers target course co-ordinators as pharmaceutical companies target doctors. Free copies of textbooks are a standard part of the deal. So too are teaching aids that go with the textbook: free CDs, websites, innumerable exercises. All of our samples are carefully packaged to be easily taught. Authors make every effort to be 'clear' (explicit, rational, orderly), to make transmission of the content as simple as possible.

This strategy may appeal to some teachers, who have little to do if they follow the textbook. They become managers of the machine rather than teachers. They are de-skilled, treated as technicians. Complex decision and thought are made redundant. Their reward would be to be paid the same for less thought and work – unless their academic managers adjust to the tactic and make them work harder (teach/process more students) for the same pay. In this way the academic managers would get more product for their money, and make life harder for teachers who want to think for themselves. Win-win for the Machine.

Students are treated in the same way, with similar effects. When this commercial model is adopted, it pre-empts issues and discussions of pedagogy. The machine will only work with a linear transmission model of education. The content, framed in linear terms, becomes the primary goal of the process. The de-skilling of the teachers and the taught is not presented as a goal of the machine. On the contrary, the rhetoric surrounding the sales pitch says the opposite. However, a management education process organised around the management textbook industry cannot easily operate with any other pedagogical model.

We did not have the resources to study management education in Mexico or Brazil, beyond determining that the main textbooks of the main US-based publishers are represented in both systems. They are translated into Spanish and Portuguese, but Latin American students are expected to be fluent in English, the language of global business.

A study of management education in the USA by Ken Ehrensal (2001) showed a similar pattern there in the Neo-Liberal heartland. Ehrensal noted the same strong dependence on textbooks, with little reference to works outside it by teachers or students. The content, he said, is highly uniform. They carefully introduce a new specialist vocabulary, so that students will learn to talk the talk. He commented on the abundant teacher resources, so that teachers do not need to know more than students in preparing classes. He found many examples from the 'real world', but so thoroughly worked over that they present a world of 'simplified certainty' in which every thing the textbook tells them to do is unproblematically right (2001:104).

Ehrensal criticises this dominant model of management education for reasons which initially seem different from Mintzberg's or Bennis and O'Toole's. They want management education to reflect the reality of the workplace. He laments the fact that it does, for a different idea of that workplace. His position is hard-line, on the left of Critical Management Studies.

For him, all these features illustrate a 'hidden curriculum', whose real content reduces prospective managers to docile members of whatever organisation they go on to join, to ensure that 'when graduates join organisations after the completion of their studies, they will accept the system of authority as legitimate' (2001:102).

The idea of a 'hidden curriculum' is a powerful one. We do not simply dismiss the idea. But is there no more to say? Is there no place for the Larrikin Principle?

Resistance is futile – isn't it?

Our version of a larrikin curriculum uses wisdom from below. We make extensive use of popular film and TV. Ehrensal (2001:108) finds that most mainstream textbooks do this too, using commercial teaching packages and popular film and TV.

He rejects both equally. We see a great difference. Commercial AV training materials for business students, in our experience, are packaged and trimmed down to fit a single 'message' and ideology. They are hard to re-engineer for critical teaching. Popular film is another matter. This is another of the Master's tools which he never owned, and still does not know how to use. Parker (2002) devotes a whole chapter to popular culture as a source of critical reflection. We agree. For instance, one of Gabriela's students devoted the journal he wrote for her course to the TV and film series *Star Trek*.

So we kick off this section with the signature phrase of the Borg, a recurring enemy for Jean-Luc Picard, Captain of the *Enterprise* in *Star Trek* 'Next Generation'. 'Borg' comes from Cyb-org, Cybernetic Organism. The Borg combines human and machine in a package ruled by machines. Members of the collective are cybernetically linked to act as one, responding to central command. All individuality is lost. Larrikins are the antithesis of the Borg.

Whenever this huge, seemingly invincible entity approaches another human or community, it announces, in a synthesised mechanical voice, 'Resistance is futile'. It tries to absorb the organic/human into the non-human collective. But every time, they are successfully resisted by the crew of the *Enterprise,* led by the autocratic Captain Picard.

Resistance is not futile in *Star Trek*. The Borg sound like ridiculous, rigid control freaks lacking the resources that come from being human, and unable to overcome their contradictions. Likewise, the Managerial Education Machine is not homogenous. It is crossed by the fissures of the Ideological Complex. It is designed to incorporate and control opposition, but the contradictions remain, becoming sites and opportunities for resistance.

For instance, not all students love the Machine. Many of the best reject it when it presents itself too directly. Andrew spoke for many:

> When I was a student here, to me university is a place and you are there to learn and also ... to express your own opinion without being turned down. Whereas on the very first day and on my very first lecture my lecturer said to me "I do not care what you think. I want to know what you have read". This was the very first thing that this person said to me, and I just sat there thinking to myself "This goes against everything and my background", when I like to say how I feel. But I realised I was in Australia, which is very much laid back here and 'she'll be right', and 'I don't care'. But then you have that sort of thing.

Like many of our students, Andrew enrolled in management and already had ideas on the balance between 'School' and 'Business' that Clegg talked of. Andrew's idea of 'business' did not cancel ideals he had concerning critical activity, thinking for himself. In his case he explicitly used the Australian larrikin ethos to reinforce his resistance. Even without it, he would surely have seen through the crude dogmatism of his lecturer.

None of our management textbooks says anything as crude as this Borg-like lecturer. On the contrary, the Management Ideological Complex celebrates the critical faculty at the very time it is producing docile students. We need to understand how and why this contradiction works, in theory and practice. The role of the Larrikin Principle here is complex. Here, as elsewhere, we do not emphasise how unique the Larrikin Principle is, as an idea and a method. The Larrikin Principle is more like a cybernetic loop. Its input may come from Mainstream or Critical Management in various forms. Its output may incorporate and shift ideas and practices of both.

To illustrate, we will use a larrikin gaze (critical, grounded, egalitarian) on one of our management textbooks: *Management: an Australasian perspective* by Davidson and Griffin (2003). It is typical in many ways: 824 big pages, (21.5x25.5 cm.), weighing in at 1.5 kilos, costing USD 120.30 (6.8 pages per dollar). There are countless bullet-points, lists and tables, and questions to test whether students have read and understood the points. It is a typical example of the Management Education Machine.

But it also has larrikin-friendly elements. The cover has a simple, childlike drawing of a man in an ill-fitting suit, balancing on a monocycle on a tightrope. His head is unnaturally small, his mouth wide-open. He is juggling five balls. Two more are sitting below, on the tightrope – waiting for him to pick them up? And drop all the other balls? This larrikin cover 'takes the mickey' out of management. It presents Management not as science but as a joke. This over-stressed manager is not laughing.

The textbook has other mildly larrikin elements. It takes topical cases from business journalism to show things going wrong in Australian and US business life. That is a start, from a larrikin point of view. It may be sanitised, but at least it connects with reality. There are also boxes headed 'critical thinking'. This is larrikinism in a box, but still allows the larrikin quality of making critical connections.

Our larrikin gaze picks up many larrikin motifs, though these authors do not emphasize them. There is a hidden need for a larrikin factor, buried and under firm control, but still present. The Management Education Machine has two 'hidden curricula' not just one. One we call the 'zombie loop'. Counteracting it is the larrikin loop.

2. Towards a critical pedagogy

'Theory' tends to be viewed with ambivalence by management students, being seen as dry, abstract knowledge opposed to the truth of practice. It is true that 'theory' can be packaged in oppressive ways, which we will try to avoid. Yet, in the case we are dealing with, management in education and in the world of business, we need an alternative to the binary of theory/practice. In this third space, ideas need to be able to circulate freely, regarding both theory and practice, in education and business alike. In this section we develop the term 'critical pedagogy' to frame that space.

Management vocabulary and its taboos

One area of life controlled by the Machine is language. As Ehrensal observed, the Management Education Machine trains students to learn the right words, to talk the right way. Behind this is a belief in the power of language to control thought in a way similar to that which George Orwell portrayed in his novel *1984* (1949). In Orwell's book, regimes devoted massive resources to controlling every area of life, exercised crucially through the control of language. This control was achieved at the price of productivity.

Orwell's target was the will to achieve total control by governments on both sides of the Iron Curtain, but his critique applies to every system which aims at total control. We will apply it to the Management Education Machine, and to the fractal series of machines in which it is embedded, upwards to world capitalism, and downwards to administrative units within firms and other organisations.

Where Ehrensal talked of learning new vocabulary, Orwell focused on exclusions. Totalitarian regimes train people not to use the wrong words, believing that if the word is banned, the thought will become impossible. There is a powerful nexus between words and thoughts, but the relationship is non-linear, with unpredictable outcomes. The Big Brother theory of language is a fantasy of the powerful.

The Larrikin Principle allows everyday terms which are prohibited by the Machine. For instance, in writing this book, we were under pressure not to write 'bullshit' in an academic text. After much thought, we finally decided: when writing about management, how could we not?

We use another set of terms equally forbidden by the Management Education Machine in our pedagogy: terms from what Parker calls 'big theory'. These terms seem to come from non-Larrikin territory, from the abstract, prestigious academic realm. Yet they are equally strongly policed by Management. We do not ignore Parker's warning, but we do use larrikin tactics. We take over these tools, which the Master claims as his, and use them freely in front of students and colleagues.

One such 'wrong word' is *ideology*. One of our 12 textbooks mentions it briefly. The rest demonstrate how the world can and should be understood without it. The French Marxist Louis Althusser had some useful things to say to both students and business about how ideology works (1971). We will feed his ideas through a larrikin/chaos theory loop, in order to say something about the Management Education Machine.

Althusser saw ideology as a machine that was much like the Borg. It took individuals and inserted them into their place in the overall structure, so powerfully that 'resistance was futile'. As a cybernetic machine, however, it was more complex than the Borg:

> The reproduction of labour power requires not only a reproduction of its skills, but also, at the same time, a reproduction of its submission to the rules of the established order, i.e. a reproduction of submission to the ruling ideology for workers, and a reproduction of the ability to manipulate the ruling ideology correctly for the agents of exploitation and repression, so that they, too, will provide for the domination of the ruling class 'in words'.
> (1971:132–3)

Unlike most writers on ideology, Althusser thought cybernetically. The task he supposes this machine can accomplish is complex and ambitious. The Machine has to divide a population into two broad classes, rulers and ruled, along the lines of Marxist theory and Capitalist and managerial practice.

But this machine keeps on making this division, applied now to all white-collar workers and management students across their lifecycles. Ehrensal saw the ideological task as simpler, to hook all white-collar workers into submission to the system. Althusser's machine is more complex than that.

The same machine has to produce managers at all levels, as well as workers. Some must be zombies who identify with executives and owners. Others, a small proportion, have to actually run things well, or there will be no profits. The rest, as Althusser insists, must have skills and competence, even if they actually run very little: in management terms, middle and operational management. The machine has to produce difference and contradiction, and also disguise the fact that these exist.

Althusser provides a systematic analysis of the Management Education Machine as if it were the Borg, and resistance is futile. Paradoxically this French Marxist provides a better guide to how this machine works than do most technicians of the Management Education Machine, who do not talk about it as a machine or explain how it works. Althusser and ideology warrant a place in a good management curriculum.

Tensions in the Managerial Complex

Tensions in the Machine show in a number of places. For instance, courses and Schools called 'Management' would often be better termed 'Business' or 'Organisation Studies'. The other two words are used, but 'Management' is over-used, for good reasons. As a term, it obscures the ambiguity surrounding where the line falls between managers and managed, by including everyone who takes a course on management on the side of 'managers'. As Ehrensal noted, this constructs their identity as rulers, whatever the reality of their situation.

The textbooks have to do work to mystify the situation, to construct a broad and contradictory concept of 'management' as an instrument of the Ideological Complex. They slip and slide across the contradiction so smoothly and so often that students may cease to notice this. When they stop noticing it, the Machine has achieved its aim. The Larrikin Principle is a counter-element, to stop that happening. Prospective rulers are meant to notice, too, but Larrikins are not designated rulers.

As one instance, Davidson and Griffin (2003) begin with a one-page section, 'Setting the scene'. This talks about CEOs as though they are the club all students will join shortly after passing the course. By the end of this page, we reach the other pole:

> The tasks of management exist at all levels of the organisation. It is not just CEOs on enormous packages who attract our attention. They may represent the pinnacle of the profession of management, but we must look at management in all its forms, at all levels and considering all applications in order to understand and improve our knowledge and skills as managers. (2003:3)

'Management' here covers the whole organisation, but we notice that 'we' (students and writers of this textbook) only need to 'look' at these lower forms, not stay in them for life. But by the next page, the democratic rhetoric is ratcheted even higher. First on a list of exemplary managers comes the 'efficient, effective, supportive and creative boss'. Two examples later comes 'the single mother with pre-schoolers, as she juggles the competing tasks of child care, home responsibilities and her education for a second career' (2003:4). Needless to say, this woman and her managerial achievements disappear immediately from the book, never to appear again.

Althusser's phrase, 'provide for the domination of the ruling class "in words"', is not user-friendly, but it captures a key paradox of the system. The machine cannot work unless it is a Trickster, run by Tricksters, skilled manipulators of words, able to see through the tricks of others, including those of their own masters. Yet these Tricksters have to be controlled, with as many extra loops as it takes.

Not all Tricksters are larrikins, but they belong to the same family of terms, and serve some similar functions in complex organisations. Althusser's Apparatus is a cybernetic model which explains the contradictions and complications of Management Education better than the simpler Ehrensal model. Althusser describes a Complex Adaptive System designed to reproduce the dominant order perfectly in new generations, absorbing elements from a hostile context.

When we frame his ideas in terms of chaos theory, we note that this Machine is a Prigoginean Dissipative Structure, an open, dynamic (and constantly changing) form of order at the edge of chaos. Zadeh's Principle of Incompatibility warns us to look at how this machine is constituted and described. The more exact its parameters, the more perfectly it tries to reproduce the previous order, the more likely such a system is to become meaningless or irrelevant.

This need for complexity is the reason why, we suggest, management has its two contradictory loops, its two 'hidden curricula': the zombie loop and the larrikin loop. The exact balance of the two will vary, from class to class, university to university, country to country, at different times and in different conditions.

These differences will have a huge impact on experience and practice. Yet, some things will be constant. The machine must have both a zombie loop and a larrikin loop, to produce and control the larrikin counterbalance. Larrikins, in this sense, are structural. Management needs an Ideological Complex that incorporates this with other contradictions.

It then needs a further larrikin loop, plugged into Top Management, to see through 'bullshit', to disambiguate the reigning Ideological Complex, and let things happen. Finally, it needs to deny it all. No wonder this reality is so hard to understand or teach, and why practice so often goes wrong.

Critical pedagogy and management studies

It is not our larrikin way to just be negative. Nor do we want to be dogmatic and doctrinaire, proposing yet another new One True Way to replace all the others. On the contrary, we listen and learn, and weave different perspectives together. We ground our analyses in experience, both our own and others, both from below and from above.

Mainstream voices like Mintzberg, Bennis and O'Toole have offered critiques which we respect. Critical Management writers like Clegg likewise contribute to the debate. Hugh Willmott, an editor of the 1992 Critical Management Manifesto, also wrote on the implications for management education. He criticised the general neglect of 'innovative' forms of teaching in management education. He advocated a 'new paradigm', drawing on ideas from outside management, captured in the formula 'a critical action learning approach to management education' (1994:105).

One strand in Willmott's proposal is the influential concept of 'critical pedagogy', associated with a Brazilian, Paolo Freire. In this section we will set out our account of this concept in Freire's life and context, to give a more complex, dynamic picture of where the concept came from, and how it may be used.

Freire was born in 1921 in rural Brazil, itself on the edge of the world system as defined from America and Europe. Yet, his ideas have made a huge impact throughout the world. This is no accident. Globalisation was integral to problems and solutions alike. His ideas did not only grow from native Brazilian soil. European ideas played a crucial role in his development, especially Rousseau, Hegel, Marx and Sartre. He was also influenced by post-colonial writers such as black Algerian Franz Fanon. When his ideas returned to the metropolitan centre, they were in some ways simply returning to their origins.

But in the process, they had been transformed. The three contexts, Brazil, Europe and African experiences of colonialism, formed a 3-body system which produced something new and different, a powerful mix, Brazilian and not only Brazilian. Global flows can be transformative, in spite of the often grubby motives of their initiators. They can change ideas as well as commodities.

Global flows impacted directly on Freire at two points in his life. He was born into a middle-class family, who were thrust into poverty by the worldwide impact on Brazil of the 1929 Great Depression. Freire never forgot the experience. He studied at university, and in that sense rejoined the middle classes, but never forgot his commitment to the poor.

This became a quality he shared with larrikins. He was a mediator, deeply concerned with social justice, mediating between haves and have-nots. He became a manager and 'change-agent' as Director of Education in Sao Paolo, designing literacy programs for the poor and illiterate. In his first venture at literacy education, 300 illiterate sugar-cane workers were taught to read and write in 45 days.

This dramatically vindicated his method's capacity to fulfil two aims at the same time: teaching literacy competence, while raising levels of consciousness. This inverts the Management Education Machine's attributes, to induce docility without producing competent managers.

Brazil then provided another twist to 3-body analysis. Freire's career as Director of Education was cut short by a military coup in 1964, and he was forced into exile. He found refuge and support in Latin America, including Mexico, where his ideas were strongly taken up. He was also influential in the USA, where he became a Fellow at Harvard University. The adverse event was turned into a positive. It pushed him out of Brazil and onto an international stage, more influential as a displaced transnational citizen than he had been as a Brazilian educator.

His ideas on education started out from a critique of the dominant education system in Brazil in his day:

> The teacher talks about reality as if it were motionless, static, compartmentalized and predictable. Or else he expounds on a topic completely alien to the existential existence of the students. His task is to 'fill' the students with the contents of his narration. (1972:45)

This is the philosophy of teaching which shapes the Management Education Machine, whose main difference from the shabby Brazilian class-rooms Freire knew is better decor.

In an illuminating metaphor, he called this the 'banking concept' of education:

In the banking concept of education, knowledge is a gift bestowed by those who consider themselves knowledgeable upon those they consider without knowledge... The teacher presents himself to students as their necessary opposite; by considering their ignorance absolute, he justifies his own existence. (1972:46).

This image captures the delusion that drives dutiful lecturers in the Management Education Machine, that loss of power is somehow a form of power. This is not just a metaphor. The same principles operate in two areas of life, education and business. Education is understood as if it were simply another manufacturing process. Behind the metaphor is an Althusserian Ideological Apparatus. In it, the same principle produces similar fractals in these different spheres of life.

Freire's methods were not designed for Business Schools, but they do offer solutions that address the core problems of the Management Education Machine. We summarise his main ideas, as potential counter-loops to the Machine: Problem-solving pedagogy

> In problem-solving education, men develop their power to perceive critically the way they exist in the world with which and in which they find themselves; they come to see the world not as a static reality, but as a reality in process, in transformation'. (1972:5)

The starting point for this way of learning is felt problems, not solutions offered for a time when the problems may come, as is common in Management Education.

Learning beyond what is known

Freire used a difficult concept, 'untested feasibilities' (1972:85) to tackle the key contradiction facing him as a teacher of uneducated peasants, committed to working with what they know while leading them beyond it. These possibilities are outside their experience, 'untested', yet can be imagined, as problems and solutions. Freire's concept is unclear and untranslatable because it is so fuzzy, yoking together contradictions. This reality is real, yet not quite real as it moves between being and becoming.

Generative words and themes

In his literacy education, Freire used the concepts of the 'generative word' and the 'generative theme' as a focus. This inverts the Machine's strategy of teaching business students a new vocabulary to replace the old. In Freire, the starting point is words the students found important in their initial reading, often from scientists or bureaucrats, which they follow up wherever they lead to: the Master's tools, to be used against him. They are 'generative' because this search can continue for ever, going behind words in common use to ideas they mobilize and realities they refer to. They reverse the Orwellian direction, where rulers control language to control the thought of the ruled. In analysing 'generative words', students become masters of the words used to control them. Unlike the rulers in Althusser's machine, who must learn to 'rule "through words"', Freire's method develops that particular skill in everyone, from bottom to top.

Freire was concerned about the problems of illiterate peasants; but paradoxically, his methods are equally suitable for highly literate citizens of the electronic era. Coronado and Hodge (2004) showed how 'generative' words and themes provide a focus for critical web research for business students in Australia and Mexico.

3. Learning from where you are

The Management Education Machine assumes the worst of students and teachers. As in Freire's banking model, students are supposed to be empty containers, to be filled with knowledge by technician-teachers. Whatever experience they bring with them is assumed irrelevant, or worse. By acting as if it were so, the Machine helps construct students in this image, with this identity. It coerces them into passivity.

Freire's method assumes this is wrong. He claims that teachers and students both have important and relevant knowledge and understanding to bring to the pedagogical encounter. But as larrikins, we do not accept this proposition without asking hard questions. Are business students and teachers really like this? And, even if Freire is basically right, how can this translate into business classrooms throughout the Neo-Liberal realm?

What are students really like?

One key strategy in addressing this question was ethnography. As teachers and researchers alike, we needed to be ethnographers: establishing trust, hearing and exploring a range of views, going behind what was said and done to recover the core values that animated them.

Students, likewise, come from many cultures, and they too can become ethnographers of the strange culture of Management Schools. We see them as 'Everyday Ethnographers' (Coronado 2008). The better we support them in doing this, the better they will be able to cope with the cultural attributes of whatever organisation they later join. In this kind of exchange, we will understand them better, and they will learn to imaginatively enter, as equals, into the culture and worldview of their teachers and future bosses.

They were our students because, supposedly, they wanted to become managers. But what did that really mean for them? What were they like when we scratched the surface? Our informal ethnography found a complex picture. Most of our students were not the slick, well-formed offspring of wealthy dynasties, studying business so they could slip smoothly into their CEO parent's place. Many were the first in their family to go to university.

They often expressed fantasies or fears that they were aiming for some top, from which lofty height they would screw all people like their parents, without compassion or remorse. But this was only something to say, for its shock value. They came to us with their values mostly intact, though open to negotiation.

Fernanda carried out a survey of students from one of her courses to see what they thought about ethics (2008). Almost all participants (95%) shared the view that 'it is important for future managers to study ethics'. Most (84%) said they had benefited from studying ethics in her unit.

Many students offered pragmatic justifications for this interest in ethics:

> Yes. Today's society expects more from their business/industry. Clean/low emissions, green practices, etc. Shareholders and stakeholders want to be seen as good corporate citizens also'.

Others saw the justification as more basic:

> Yes, as well as forming a personal basis of ethics, it is important for managers to take into consideration ethical practices and not just profit and finances. Companies in the future need to be ethically responsible.'

For others, ethics had a vital connection with leadership:

> [Ethics is] important because as a manager you are at times looked upon by those under you as a role model and therefore by your demonstrating ethical behaviour, they can learn from you.'

Across this cohort, there was an impressive commonality of viewpoint. For them ethics was not opposed to good business behaviour. Ethics was not just for lowly workers or an easily duped populace. Everyone should be ethical, responsible citizens. This was not just a minority viewpoint. 95% is a stunning figure. If this is what our students think, it would be a terrible betrayal if our management course taught this attitude out of them, as Ghoshal feared.

In other conversations we found that most thought it would be hard to keep their core values once they joined an organisation. But that was not because they thought they would change these values. What they feared was that their employers would make them untenable, that the world out there was a jungle, where only ruthless exploiters thrive and survive, where the only rule was to maximise profit by any means. They believed that this was the world they were heading for, but many did not like it. Gordon Gecko, in the film *Wall Street,* was a nightmare, not a role model.

They saw their first few years of employment as a time of weakness and vulnerability, where they would have no chance of transforming their workplace, much less the world. Yet the best of them, the most promising future managers, wanted to know in practical detail how to change the world. We had all taught anthropology and sociology students who knew how to talk the radical talk. We were impressed by the quiet determination of some of our management students to make a difference in practice.

Nor did they lack relevant and real experience. All were doing paid work before and during their courses. This was too menial to count as relevant experience for Mintzberg, Bennis and O'Toole. Yet this is experience from below, larrikin knowledge. It gives a valuable perspective on organisational life, good and bad management.

For instance, Allan reflected on his work experience:

> I find that also with a lot of my friends who work in different organisations the people they work in like fellow employees are sort of centred around you know: 'we don't need management to tell us what to do, we don't need someone being autonomous [autocratic?] or what not and saying, this is the way you have to do it'. They think that, yeah we think that we're intelligent enough and responsible enough to work by ourselves and not have someone constantly on our backs. Which is I think the majority of people. If you provide a good working environment and you just left the right people with the right organisation then that's the way it should work.

Allan uses here a version of Freire's 'untested feasibilities' to extend his experience. He uses his peer network to do so. The lessons he is learning may not be what the Machine would like him to learn, but that is another matter.

Three-body Critical Pedagogy

We also used ideas from chaos and complexity to bridge abstract theory, management practice and personal experience. For instance, Gabriela adapted Poincaré's three-body systems to her course *Business, Society and Policy*.

The terms of this course already have a three-body form, yet students and textbooks often reframe it as a binary. Business is seen as the dominant reference point, opponent or ally of government, which regulates too much, as Neo-Liberalism says, or is justified in some of its practices. 'Society' is so nebulous that it tends to disappear from discussions.

In her teaching, Gabriela formed groups of three students around the various issues contained in the course, one for each of the three terms, government, business and society. Students were given no choice regarding which one they would represent.

This could have risked promoting the kind of amorality that Ghoshal objected to, a capacity to argue on behalf of any case their employer told them to. In practice it served to push them out of their comfort zones, to try out the 'untested feasibilities' of the range of positions that exist on any complex issue. Rather than the amoral conclusion, that truth does not matter but is exactly what employers say it is, they learnt the complementary lesson. Truth does matter, precisely because no one version is the sole Truth.

Three-body analysis can also help connect management principles in textbooks with the everyday experience of being a student. For instance, the following is Hitt et al. on how to develop an action plan:

> **Sequence and timing.** A key element of an effective action plan is the sequence and timing of the specific steps or actions that must be taken. One of the common tools used to graphically display sequence and timing of the specific action is the Gantt chart...
>
> **Accountability** The second key aspect of effective action plans is a specification of who is accountable for which specific actions...
>
> However, no matter how carefully you work out the implementation plan, you will still need to monitor and adjust your implementation efforts, because unanticipated events almost always happen (Hitt et al. 2007:280–1).

This describes linear analysis, supported by graphic technology. Gantt diagrams introduce one dimension of real-world complexity: the fact that actions do not form a single chain, but often overlap. But this modification towards complexity will not be enough. The beautiful machine will hit chaos, and Murphy's Law will strike.

Students do not have to wait to become managers to have relevant experience to understand for themselves the real-life problems of over-linear planning. The next text comes from our corpus, from an administrative staff member:

> Sarah:
>
> A student needed to lodge a Form and the information on the Form wasn't what the people who were accepting the Form wanted, you see. And the student came to me and said that they wouldn't accept the Form because of a technical reason. And the reason there was a problem was because the Head of School had failed to complete a task that he said he was going to do. He just didn't get around to doing it, and that had meant that other things didn't get done, which meant that the Student Centre wasn't going to accept the Form.
>
> So the student came to me and said, "What will I do?", so I got on the 'phone to, everyone knows WHO, at the Student Centre, and I guess you sort of try to engender a relationship that might not really be there. What I did was I called her by her first name, and I passed a little comment about it being a difficult time of semester, or something in the way of a joke or something that recognises where she is coming from, or what her frame of mind might be, and then I presented to her the situation: that the student came to me, the student couldn't get her form accepted and I explained to this lady in the Student Centre what the problem was and why University processes hadn't been put in place.

> And she said to me, "Well, OK. If you write a note on the Form requesting Enrolments to accept the Form, if you make a note on the Form and explain why there is a problem, and mention that the School was going to do something, then I will accept the Form." So I wrote the note on the Form, the student took it up to the Student Centre and the Form got lodged and accepted.
>
> In the meantime I did send an email to someone else to say "Can you please get the Head of School to do what he said he was going to do", so I had to push that along in that direction, but at the same time get the Form lodged.

When students do a Gantt diagram of this process, they get a vivid image of the unintended complexity created by everyday mistakes, the typical complexities of organisational life.

The biggest fallacy is the impression Hitt gives that the Manager and the plan are fine, others are the problem. In this typical case, Sarah goes to great lengths to make the system work as it is meant to. She uses interpersonal skills and larrikina strategies to get a key other person on-side, not for her own benefit but to save the student and the system. She is not the problem but the solution, to a degree that cannot be glimpsed in the pages of management textbooks.

Students are the third body in a system whose other members are management systems operating from above, and administrative structures at other levels. We quoted Andrew before as a larrikin student. Here he comments on the system:

> With my enrolment a few years ago I had an issue with my forms. I had signed off on them and everything was fine but they had been misplaced, so I had to do them four times over, and every time it got lost. And then I was threatened with disenrollment, because I hadn't filled out the forms, even though I had. So that is when I went above their head and sent a very strongly worded email to all and to whom it concerned, not only their own boss but mine, and a few other people that were involved with the whole thing. And lo and behold, within two hours I had a resolution.
>
> So that was an example, and it really showed that you have to keep fighting. You can't just say "Oh yeah, whatever". And keep records and keep receipts, very important.

The problem for management teaching with this example is not that students lack the experience to understand such foul-ups, but the contrary. The problem is that this knowledge interferes with the managerialist fantasy that management control operates with seamless perfection.

Andrew analysed this real-life system pragmatically and effectively. He was one body (student) in a three-body system, and also incorporated all three elements. He is effective because he understands how the situation looks from above, management, from the side, administration, and from below, as a student. This internalised three-body system as a fractal was a resource.

One take-home message for him was to avoid a casual larrikin style. In order to have larrikin effectiveness in taking no nonsense from authorities, he had to use their own weapon of precision: carefully kept records and receipts. This contradiction makes our larger point. Mainstream practices and larrikin alternatives alike are constituted by contradictions. There is no single, simple formula for dealing with the complexities of organisational life.

What should management teachers know?

Management teachers constitute the other part of the management-teaching equation. They can be ranged along a continuum between two poles. In linear transmission models, teachers should know much more than students, especially at the beginning. Knowledge should be transmitted in regular steps. In principle, the management curriculum could be assessed through critical path analysis. In mainstream management education, these assumptions seem like common sense.

At the other extreme, Freire's form of Critical Pedagogy has a different role for teachers, and requires different ways of knowing. Far from teachers having to know more than students, and emphasise the knowledge they know more of, Critical Pedagogy shifts the balance towards what students already know, and builds on it.

Bennis and O'Toole hold a complex position on the disciplinary range needed by management teachers and researchers. On the one hand, they want a greater range, including social sciences and 'scientific' management. They also lament the number of professors 'with limited real-life experience'.

We will use an ethnographic gaze on our own situation, in our university, to open up some of the complexities of this issue. Recently our College introduced a rule requiring all part-time tutors to have industry experience. In doing so, they reflected the concerns of such senior management educators as Mintzberg, Bennis and O'Toole.

Our team is acutely aware of one problem with this policy. None of us has worked extensively outside the university sector, nor are we trained in 'management'. We come from sociology, anthropology, cultural studies, semiotics, and chaos theory. We apply these disciplines to problems and issues of management and organisation, and draw our strength from our interdisciplinary background. Much management and organisation theory and research originates in social science disciplines.

Our school, with its new staffing policy, could not employ us as tutors. We are aberrations. Yet, we were employed to be the aberrations we obviously are. There has been no move to sack any of us, but the contrary. We have been asked to develop new units in our areas of expertise. We are valued members of staff, teaching and developing successful courses, publishing in good journals.

The paradox comes from contradictions inherent in management education. In a *School* of Business, we have a role. It is only a School of *Business* which would find us a problem. And in this case, create one. We can no longer employ good sociologists without industry experience as tutors.

This contradiction in the organisation mirrors contradictions of our own which make us better ethnographers, as well as potential contributors to our School of Management. Our background includes journeys we have made, between where we have been and where we are now.

As we look we see the extent to which we have always belonged to organisations. We were all involved in one way or another with unions, blue-collar as well as white. We have all taken on management roles within our universities, including directors of units, heads of departments and programs, Faculty Dean. This is experience on both sides of the divide which too often exists between 'Management' and 'Workers'. This double identity is basic in a larrikin.

All academics in all Faculties have relevant experience of management and organisations. This experience does not count, neither for our College nor for the world of Business Education. But Schools of Management are of course organisations. So are universities. So is a university system.

All university staff work in an organisation. So do all students, in the same organisation. But we and they are trained not to think of it as an organisation, nor to see its processes as examples of management. Because they do not do so, we all lose access to a rich set of common problems and experiences that could give a firm smack of reality to the study of management issues.

The scope of 'Management'

Our histories and fissured identities, as teachers and university 'managers', point to a split with wider ramifications. There is an analogous split in the scope of 'management' in our textbooks, which is an aspect of its Ideological Complex. 'Management' is either a generic attribute (which attracts large numbers of students who will never be CEOs) or it is the specialised function and role of 'top managers'.

This split is reflected in ambiguities about what kind of organisation management degrees equip students to manage. Managerialism claims management is a generic attribute, applicable to all organisations. However, most examples in our textbooks refer to business organisations. That becomes the taken-for-granted reference point.

Only three of our 12 textbooks (25%) even mention not-for-profit organisations, but two (Hubbard 2004, Robbins 2006) give the matter serious consideration, which throws light on the issues the others are silent about. Hitt et al. (2007) do not have that category but sometimes use university organisations as examples, mostly of what not to do.

Robbins' (2006) discussion reveals complexities that may underlie the silences of the others. They discuss it under the question: 'Is the manager's job universal?' More specifically: 'Is the manager's job the same in profit and not-for-profit organisations?' Their answer is definite: 'For the most part, yes' (2006:19–20).

They then attack three 'myths', all negative judgments about not-for-profit management compared to proper 'businesses': that they are more political, more tied up in bureaucracy, and less efficient. No, no and no, they insist.

Yet their text implies the opposite judgement. In the rest of the book they refer to not-for-profit organisations on only two occasions. Both illustrate the myths. Otherwise, like their fellow textbooks, they do not mention not-for-profit organisations.

This confusion is so brief that it may hardly be noticed, but it reflects a core contradiction in the managerialist complex, between the universal scope of the management function, which Robbins et al. insist on, and the privileging of forms of management in business. This is what splits us from our proper management colleagues, yet has us in the same school. The same principle will split all our students who go to work in government, non-government or not-for-profit organisation. Since this will be more than half our cohort, this is an important split not to talk about, as most textbooks do not.

What we do in this section can be called auto-ethnography (Duarte and Hodge 2007). We use our insider-outside status to explore the culture's lived contradictions, often otherwise not stated, concealed from self or others. This is often the best or only method to explore such important but hidden complex meanings.

This comes out as larrikin behaviour. We cross many boundaries by talking about our own institution in public. We seemingly show a lack of respect to our bosses, who made and enforce these rules. But we broke no contract of employment or rules of ethics. This is not confidential data. It is only what we have seen with our own eyes and reflected on.

Yet, as we insisted at the beginning, larrikins respect some constraints. These are people we work with, whether bosses, colleagues or students. We have no wish to make them look bad, or break trust. If we gently point out contradictions, in a positive spirit, that is fine according to our larrikin code.

Making sense of the big picture

We finish with a final paradox about Management Education: Neo-Liberalism is everywhere, yet not named or studied critically. Local ethnography is not enough to understand properly even where we are. Every local level nestles in larger structures, which need different disciplines to unpack them. Clegg lamented what he saw as the dominance of economics in business education, but this dominance is even more effective because it is almost invisible within the management curriculum itself.

Later in the book we look at ways Management Education talks at great length about 'globalisation' as the indispensible context in which to understand all forms of organisation today, while neglecting the key tools needed to analyse macro- and micro-structures and their complex relationships. We look at some theorists of Neo-Liberalism, including Nobel prize winner Milton Friedman, and modern critics of Neo-Liberalism like another Nobel Prize winner, Joseph Stiglitz. Friedman is mentioned in only two of our textbooks. Clearly, winning a Nobel Prize in Economics does not cut it in Management circles. Not even being a founding father of Neo-Liberalism.

All Australian universities, our own included, are object lessons in Neo-Liberalism which are never discussed in those terms in our management courses. Most lecturers and researchers feel the pressures, but do not join up the dots.

As one instance, our university has been undergoing 'restructuring' for a number of years. So has every other Australian university, government department and major corporation. It is a global disease in the English-speaking world and beyond. This produces many adverse changes that students complain about; cancelled courses, sacked staff, excessively large classes, lack of support staff. Yet this process has no place in the official curriculum of universities like ours.

The common pattern across the global tertiary sector is driven by Neo-Liberal principles. These force university executives to manage these institutions as if they were private corporations, following the dictates of Economic Rationalism, as Neo-Liberalism is often called in Australia (Pusey 1991).

In Australia as elsewhere, the application of this model is rigid and controversial. It has produced many dysfunctions which students notice and object to. In the words of Simon Marginson, Australia's leading specialist on higher education, 'Government has achieved the goals of fiscal reductions and export growth at the price of Australia's larger capacity as a global knowledge economy' (*Australian* 11-7-07:21).

Yet Neo-Liberalism acts in complex ways. Elsewhere Marginson says:

> First, the Western environ is established. A ring is drawn around practices, sites, and subjects, within which lies the space ('international education', 'quality assured practices', etc) that is marked out for organisation. Other

cultural and educational practices, such as those in the home countries from which the students have come, are pushed outside of the circle and ignored. Second – within the circle agents bearing 'difference' are subordinated, by defining how far they must travel to achieve sameness by eliminating educational practices that are 'deficient', i.e. habits of learning that differ from those prevailing in the country of education. (2008:20)

The reasons that lead students to move on this scale are not all good. Yet move they do, in massive numbers, creating richly multicultural spaces wherever they go, and wherever they return to. In 2007, 11% of the student population of the University of Western Sydney came from many overseas countries, too many students to ignore. Yet as Marginson says, this diversity is treated as a problem of control, not a cultural resource, by university systems in the developed world.

One reason is the ethnocentrism that continues under Neo-Liberalism. As Marginson says, opportunities created partly by Neo-Liberalism are then missed by education systems currently shaped by the Neo-Liberal linear, one-size-fits-all approach. These contradictions can be viewed more productively through a chaos and complexity framework. The contradictions of Neo-Liberalism are self-cancelling and unproductive because they are unacceptable to its linear framework. Yet if they are accepted in a non-linear larrikin perspective, then the all-important next question will arise: what can be done?

The crisis in business education

Business education proved a good place to begin this book, because it is not as disconnected from the wider world of business as some fear. On the contrary it is a fractal level containing all the elements, forces and issues students will soon confront, even if their teachers may not realise it or use it as a teaching resource. No one should worry if there is a 'crisis' in business education. Given our turbulent times, the real worry would be if there were not.

Chapter 2

The Larrikin Principle

The Larrikin Principle has no official status in management, or any other brand of theory. It would be un-larrikin to pretend otherwise. It is more productive to leave the concept open, unsystematic, accumulating an inexhaustible range of dynamic, heterogeneous ideas around it. In this book, it emerges from the intersections between culture, history and chaos theory, to engage with some core issues for management and organisation studies.

Yet exploratory ways of proceeding are scientific, as good science is often done. We illustrate with a popular science book on the search for 'dark matter' by Ken Freeman, Duffield Professor of Astronomy at the Australian National University.

Freeman tracks the curious fate of astronomer Fritz Zwicky's 1930s proof of 'dark matter' in the universe. Zwicky combined good Newtonian mathematics with Einstein's theory of relativity to demonstrate that the shape and observed movements of a galaxy cluster require large amounts of 'dark matter' to be hypothesized.

Zwicky's method was good science. He showed that the current model, based on luminous bodies, did not fit what was observed. His proof was ignored. Freeman himself 'while studying something entirely different' (Freeman and McNamara 2006:37) made a key contribution in 1970 that helped to make this theory the orthodoxy it now is. He showed mathematically that observed data did not make sense unless there was 'dark matter' in the halos of galaxies to hold them together.

Zwicky demonstrates mainstream science's need for people who produce unorthodox ideas, and stubbornly stand by them: 'an irascible fellow who liked to disagree with the deeply held scientific beliefs of his fellow scientists' (Freeman and McNamara 2006:32). This is part of the Larrikin Principle.

We will also use 'dark matter' as a metaphor to frame some basic properties of organisations. Just as galaxies and orbits do not make sense unless the neglected 'dark matter' is included in calculations, organisations do not make sense without recognising something very substantial which holds them together. This allows them to be more dynamic than they would otherwise be. This is the domain of the Larrikin Principle.

Nor is 'dark matter' a minor element, safely ignored. There may be 90–95% dark matter and related substances in the universe. Analogously, linear Managerialism has a causal model in which official rules, purposes and actions of management are primary and determining. In this picture the complex agency of 90% of the rest of the organisation is a distraction or irritation at best, not intrinsic to the dynamics of the whole.

Scientists who ignored Zwicky for 40 years had the excuse that dark matter was far away. Linear Managerialism has no such excuse. Now there is a growing body of work in organisation studies which recognises the massive role of informal systems and forces described variously as 'culture' and 'social capital', which we examine later, in Chapters 7 and 8. This is the medium in which the Larrikin Principle operates best, and is most needed.

Or to change the metaphor, dark matter can be seen as the unconscious of the system, underlying forces which explain the non-linear complexity of what actually happens. In this metaphor, the Larrikin Principle taps into the unconscious of linear management, bringing its repressed knowledge into the light. That is how we use it.

1. Larrikin practices

We begin our unpacking of the Larrikin Principle on a humble scale in the present, with auto-ethnography, using field notes collected by Bob in a visit to Mexico in 2007. Auto-ethnography is flagrantly not objective, with a sample of only one, yet it allows a usefully rich description of this lowest of fractal levels.

The incident occurred in a departure lounge in LA airport. Bob's connecting flight from LA to Mexico was delayed by 2 hours, and he was worried that Gabriela would be waiting all that time in a cold Mexican airport:

> I overheard the Mexican in a business suit sitting beside me using his mobile to ring his family to tell them he was delayed. I didn't have a functioning mobile, so I asked him if I could borrow his phone. I would pay him, of course. I felt a bit of a larrikin, breaking the unwritten conventions of airport lounges, but what would I lose? He looked taken aback at first, but agreed, punching in the numbers I told him (suspecting I might be ringing Australia for a long conversation at his expense? A control reflex?) After I rang off, I asked how much it cost. 'I don't know', he shrugged, and smiled. I thanked him. 'De nada' he insisted. (From Bob's journal, 15 June, 2007)

The incident itself is trivial, but it is a fractal of many larger systems that surround it. What is woven together in a single person's experience, recorded in an everyday story, can be highly complex, illuminating the analogous complexities of a global world.

Bob begins by reflecting on larrikinism. That is unsurprising, given that he is researching this project. But the way he talks about it is less conscious, more revealing. He sees himself as only 'a bit of a larrikin', only in this one action. That makes him a fuzzy larrikin. He combines some larrikin qualities, to a degree, with other qualities. The idea of fuzziness is a simple but powerful way of avoiding simplistic stereotypes. Bob is a fuzzy larrikin/

professor: only partly a larrikin, still partly a professor. So was Zwicky, to a different degree.

Larrikinism is associated with Australia, and Bob is Australian, but this is a transnational encounter in a quintessential transnational space, the lounge of an international airport. The larrikin move is understood by the Mexican. It is not exclusively Australian. As the notes make clear, Bob's move might not have worked, and almost did not. The fact that it did shows a creative bridging of different but congruent cultural traditions, from Bob's Australia and this man's Mexico.

According to the linear ideology of dominant management thinking based on self-interest, the Mexican's behaviour is irrational, unless it is described in terms that do not connect with reality: 'you scratch my back and I'll scratch yours'. It is not likely that his back will ever be close enough for Bob to scratch it. He could not reasonably expect to meet Bob again and cash in on his kindness. If he did gain favours from someone else at a later date, his action today would not contribute directly to it.

This behaviour does not make sense in terms of the linear concern for the 'bottom-line' that we assume this man hears everyday from business colleagues. It more likely comes from the Mexican culture of reciprocity he was raised in, which has affinities with Brazilian culture patterns, as well as with Bob's Australian culture.

Yet, this seemingly irrational behaviour has roots in an alternative economic system which was once almost universal amongst humans. Anthropologists call it a 'gift economy' (Mauss 1966). In gift economies, goods do not have a precise cash value, realised through exchange. Their value is social, holding the society together.

The benefit to individuals is less clear. They can feel good about themselves, at a tolerable cost. The Mexican in this case is offered a translation into cash value but rejects it. Presumably the return, using the cash system, is not worth having compared to gains in the non-cash system. The values on both sides of the equation are kept fuzzy. Neither knows what the other thought, but that ignorance smooths the transaction.

Later we will look in more depth at systems like this, commonly called 'informal' in the management literature. We will suggest that they are like dark matter, continually falling outside linear analysis, minimized or ignored, slipping in and out of consciousness. Yet they underpin the workings of substantial parts of the total set of interactions in any organisation, any exchange.

Defining the larrikin

The behaviour that triggers this reflection is rule-breaking, a larrikin signature. But this is only a minor rule, from an informal social rule system. It is equally a small transgression or minor creativity. This is not an attack on the rule system for its own sake, but a minimal change needed to achieve the effect. Both men see the transaction in these terms. That is probably why the Mexican obliged.

Both those involved are men. Issues of gender are part of the 'dark matter' of this exchange, but in complex ways. Gender also enters into the idea of 'dark matter' itself. English *matter,* Spanish and Portuguese *materia*, both come from Latin *mater*, mother. In this exchange, Bob recognised that the fact the Mexican was a man was factored into his decisions. He would have felt uncomfortable making the request of a woman. Yet he felt that a woman could have done it too, in a different way. Gender has effects, but they are non-linear.

One complicating factor here is trust. The Mexican begins by not trusting Bob, but then decides to do so. Without trust, nothing would have happened. Trust is a key catalyst. Something Bob does generates it, and something in the Mexican responds to it. It binds this system together briefly, long enough for good things to happen. The exchange is so minor and brief that it is easy to miss the fact that there is a system, surprisingly complex and effective. It is a dissipative system emerging at the edge of chaos, performing a range of functions before it disappears.

Its value to Bob is clear. As a result, he can relax, knowing that Gabriela will not be worried. We do not know its value, if any, to the Mexican. But it also solves a minor problem for another player. Bob's problem was created in the first place by the problem with the plane schedule. The computerised system of bookings and flight schedules is a miracle of linear processes, worthy of all the praise it receives. But inevitably small glitches happen. When problems in a linear system hit the non-linear set of systems they are embedded in, 'shit happens', to quote Murphy.

In such cases, passengers become agitated. The airline knew this, and took one typical action. A young woman at the desk was instructed to broadcast a calming message every 20 minutes. She apologised for the short delay, and promised the problem would be fixed in half an hour.

By the third iteration, this tactic was not calming Bob. He grew irritated. He classified the announcements as 'bullshit'. They decreased his trust in the airline. The Mexican man's help solved his practical problem, i.e. how to get a message to Gabriela, and improved his mood. He benefited from that, of course. So did the airline, now having one less irate passenger on their hands.

We draw a number of propositions from this minor incident, to explore further:

1. Big, linear, formal systems never function perfectly as planned. They produce disorder, dysfunction, chaos: 'dark matter'.

2. These consequences are typically managed, to some degree, by negative feedback loops in non-linear, informal systems emerging out of the 'dark matter' of the context.
3. These corrective systems form complex dissipative structures; open, dynamic, fuzzy, informal systems that creatively manage new flows and new challenges.
4. The Larrikin Principle, under that or other names, is a set of strategies for acting effectively in this kind of situation. Although it has historical associations with Australia, it is a generic attribute, found across many other cultures and periods. Although it has been cast as a masculine attribute, its relationship to gender systems is far more complex and less determined.
5. The Larrikin Principle mediates between the massive, sometimes disorderly non-linear systems of the context ('dark matter') and the dominant linear systems. It 'mends' problems created by over-elaborate, dysfunctional linear rules by bending or evading them or making them more fuzzy. It creates conditions of trust and good will which allow complex reciprocal processes to form and take place. It is a catalyst for complementary systems that resolve the otherwise difficult and intractable problems of the dominant system. It is essential to their viability.
6. In the current state of linear managerial ideologies and practice, the Larrikin Principle is treated as inconvenient or irrelevant. The larrikin function is unrecognised and unrewarded, its positive effects treated as a magic that is owned by bosses. The 'magic' would work better if it were recognised and supported.

The double face of larrikinism

Thus far we may seem to be talking as though larrikins are always a good thing. In fact, the larrikin has a double ambiguity which cannot be tidied up or removed, because it is intrinsic. We begin with another entry in Bob's field notes:

> An embarrassing story in the Mexican newspapers today. Apparently the Australian Big Brother program, facing falling ratings, ran an over-the-top Mexican-themed game show to liven up the Big Brother house. In one segment, teams wearing sombreros threw balloons filled with Tabasco sauce at a large Mexican flag, which was hung the wrong way round. The winners were those who made it wettest. An obscure allusion to 'wetbacks'? The Mexican embassy complained to the producers, and an international incident almost occurred. The producer claimed that this was only intended as 'a sincere tribute' to Mexico but in any case apologized profusely, and promised it would never happen again. This was Australian larrikinism on the international stage, and it didn't look good. (Bob's journal 22 June 2007)

It is Bob the self-styled larrikin who judges this 'embarrassing'. He initially supposes that Mexicans would misunderstand the incident and judge it harshly. However, the Mexican press generally reported the event without indignation. For instance, Juan Hernández, in the left-wing La Jornada, referred dismissively to the 'nationalistic souls inflamed by a show of extreme frivolity' (LaJ 22-6-07:4). He was especially amused at a quiz in which contestants were asked if 'Mexico has achieved its independence... from the USA'. His tone suggests he thinks this is either Australian ignorance or good satire.

Moreover, Big Brother is not a core Australian cultural institution. This global franchise originated in the Netherlands in 1999. Metastases spread through Europe, also reaching both the USA and Australia. Mexico and Brazil both have their version of Big Brother. Mexicans already know at first hand the status of the show. Globalisation had already washed Big Brother and cultural stereotypes of Mexico onto Australian shores.

Clearly, Bob was wrong about how Mexicans would react. Hernández easily saw it as a *tontería*, a piece of silliness. It shows what Australians would call larrikin behaviour, but many Australians expressed shame over it. Larrikinism is not an essential Australian quality. It is equally Australian to dislike it.

The same is true of Mexican responses. The Mexican Embassy in Australia expressed official disapproval. Hernández has a relaxed, 'larrikin' response. The Larrikin Principle does not divide Australians from Mexicans and everyone else. It divides Australians and Mexicans alike, along the same line of cleavage. It even divides individuals within themselves, as Bob shows. It is a fractal form with many levels.

Many larrikins in our corpus showed different degrees and forms of this ambivalence. Here is Denis, an Australian:

> I think the increasing bourgeois element of society in this country, it's perhaps to some extent undermining the practice of that larrikinism. That's not to say that it doesn't happen. In certain aspects of our business history, the Alan Bonds, the Skases, directors of various companies like HIH and OneTel. I think there's elements of larrikinism there as well as outright illegal practices.

For Denis, larrikinism carries radical ideals which are under threat from the 'bourgeois element' – yet it is also criminal, to a degree. His two modern larrikins were both convicted of flagrant illegality, not just 'elements' of larrikinism. It is clear the Larrikin Principle is not a simple concept.

Maggie expressed a point of view that was common among Australian women:

> Larrikin characters? I've come across a few. Everyone's got opinions of everyone else. And those larrikins, you know, everyone has an opinion of them. And you steer clear of them. Or yes they're nice at face value but for everything else you know um, not worth it. But I don't know, like I mentioned before, like I think you get that in any kind of organisation.
>
> I think that's when the politics really comes out, and so from what I've learnt is, I don't want to get involved. At all. Like, you can't remove yourself completely from the politics of any kind of organisation, but to be aware but removed, I think that would be the best bet. Because there's nothing worse than associating with these people that are looked – not really down upon – but not thought of so highly in your organisation, and you end up catching the same kind of opinion.

Gender is important in Maggie's concept of larrikinism, as with many Australian women. But what creates her acute ambivalence seems to be their reputation, as it positions them in organisations. This negative reputation is contagious. Even though she is a woman, she could 'catch it'.

But her story implies another point. Maggie clearly comes from an organisation which does not value its larrikins, and wages war on them. She does not want to be part of this war. Her strategy is to be 'aware but removed'. These are words of someone who does not feel secure in her place of work, or committed to the organisation. This organisation, which treats larrikins badly, also seems to have serious problems of morale.

Larrikins, larrikinas and the gender problem

On September 5, 2006, Steve Irwin, star of TV's The Crocodile Hunter, was killed by a Manta Ray while filming an underwater show. He was the most high-profile 'larrikin' of his day. Tributes usually mentioned the fact, as a positive:

> He wasn't known as the Crocodile Hunter for nothing. With his trademark khaki shorts, size-12 boots and his famous "Crikey" catch-cry, Steve showed no fear as he rugby-tackled, wrestled and trapped some of the most dangerous animals in the world.
>
> It was all these daring encounters, his unabashed passion, and his larrikin charisma that won the hearts of millions of people from all walks of life and from all over the world.
>
> Steve, a self-proclaimed wildlife warrior, always believed he was put on this planet for the one and only purpose of saving the environment (Gibbs 2006:17).

But 12-year-old Paige and her 10 year-old brother Chris showed cross-gender solidarity in their published tribute to Steve:

> We are so sad about the loss of Steve. We often pretended we were Steve and Mum was a croc (Daily Telegraph Sept 9, 2006:28).

In this highly charged (far-from-equilibrium) situation in the wake of Steve's death, gender played a complicating role. Feminist icon and larrikina Germaine Greer aroused a ferocious response when she criticised this icon soon after his death, pointing out contradictions between his wildlife credentials and his treatment of animals. One of her few defenders summed up the debate from this other side:

> The ferocious response of "ordinary" Australians, egged on by people such as John Birmingham ("Greer's feral attack reflects elitist conceit"), to Germaine Greer's dissenting view of Steve Irwin is shameful on many levels...
>
> 'The essence of our system is the right to dissent from common "wisdom". Howling down Greer in Australia is no different to banning secular academics in Iran.' (Dov Midalia, The Australian 2006:15)

Does Australia only accept loveable male larrikins? Would an awkward intellectual larrikina have to die first, to be acceptable?

2. Fuzzy analysis and the Larrikin Principle

Everyone who deals with situations of great complexity has to cope with the fuzziness of language. Words shift and buckle in spite of the wishes of their users. In this section we try to offer guidance and help. We illustrate fuzzy critical analysis as applied to larrikins, with the larger aim of offering analytic tools to everyone in the world of management and organisations who grapples with this kind of complexity.

i) Fuzzy logic. Thus far we have used Zadeh's key term, fuzziness, fairly loosely and informally. This is appropriate for this term, up to a point. It is a fuzzy use of the term fuzzy. Yet the power of the term comes from the fact that it contains or tolerates so many contradictions and makes them manageable and thinkable, including a degree of formalism. Zadeh wrote:

> The pervasiveness of fuzziness in human thought processes suggests that much of the logic behind human reasoning is not the traditional two-valued or even multivalued logic, but a logic with fuzzy truths, fuzzy connectives, and fuzzy rules of inference (1973:28)

Two-valued logic applied to larrikinism would consist of tight definitions, strict deductions, and certain truths. All larrikins would be identical, and different to everyone else, in terms of a specific criterion. It would be possible then to make definite, certain statements: e.g. larrikins do this or that, therefore managers do not.

As we have seen, larrikinism does not have a single meaning or value like this. Meanings are spread over a range of qualities. The value swings between strongly positive and strongly negative. We do not need Zadeh's help to see this. What he provides is a different attitude to it. He sees this quality as inherent in human thinking, an effective kind of logic in its own terms.

Importantly, he accepts fuzziness, and builds a whole edifice of thought on this basis. Traditional logic is fine – clear, definite, certain – when it can find crisp terms to apply to reality, and problems that can be described in these terms. Outside these conditions, traditional logic cannot be easily applied, and where it is it misleads. Using the dark matter analogy again: the crisp logic of linear science is unable to see much of what makes up the situation, which is invisible in its terms.

ii) Pareto analysis. To help make this idea familiar to management thinkers we will discuss it in relation to the Pareto Principle. Pareto was a 19th century Italian economist who observed, as a rule of thumb, that in the societies he studied, 20% of the population produced 80% of the wealth. In the 20th century, his principle gained new currency as adapted by Total Quality Management into its strategy for continuous improvement. In TQM analysis of faults in a given process, 20% of factors can be expected to be responsible for 80% of faults. With this guide, efforts at improvement concentrate on the 20% to get the best results with the least effort.

Pareto's principle is an instance of a mathematical tool of chaos theory, the Power Law we described in the preface. Pareto's 80/20 rule of thumb is only one of many instances of Power Laws. The general point is that in many systems, at a given time, a small proportion of cases produces a large proportion of a given effect.

Seeing larrikinism in these terms, we can presume that an organisation may get roughly 80% of its larrikin-effects from 20% of its larrikin staff. Taking this to the next fractal level down, individuals may produce as much as 80% of their larrikin effects from only 20% of their behaviours. That is, even someone who is regarded as a larrikin will only be a part-larrikin, a fuzzy larrikin who is not-larrikin 80% of the time. Conversely, larrikinism will also be found, to some degree, in most people in the organisation.

We illustrate some implications of this analysis with Martha's story. She identifies a small number of individuals as larrikins, known and targeted by managers. She avoids them in case these managers include her in the same category. That is, she fears they will extend the category 'larrikin' to include her, and act accordingly. In terms of fuzzy logic, her worry is that her fuzzy larrikinism will be fixed by them into a crisp category which they will use against her.

If we are right to suppose that her organisation is dysfunctional, and creates fear in employees like her, then probably many others would agree with her. That suggests that many others in this organisation may be critical, to a degree. If this workplace is as politicised as she implies, it may need more larrikins to speak up in order for these managers to get the message.

This example suggests why crisp logic can be so attractive to managers, and so dangerous. Its attraction is to make the problem seem simple: if 20% of the staff are 'bolshy', sack them, and 80% of the complaints will go away. But if larrikin feelings are diffused, to a degree, throughout the workforce, it cannot be eliminated in this way, short of sacking everyone. And even then the core problem would remain: the poor managerial practices which provoked larrikin resistance in the first place.

iii) Dynamic fields of meaning. Words are fuzzy because they are part of a complex, dynamic field of meanings. It is only by mapping them in a way that captures some of this complexity that we are able to understand why, and where, they are so hard to pin down and yet so potent. We illustrate this general process with the word larrikin. Crisp definitions try to fix boundaries between words. In practice, words in common usage always overlap with others. For instance, in our media corpus we found four terms which overlapped with larrikin: 'maverick', 'pioneer', 'outlaw' and 'criminal'.

One quality relevant to all is a relationship to rules and law. At one extreme are criminals who break society's laws. Outlaw seems equivalent to criminal, but has a different inflection. Outlaws not only break laws, they are also on the fringe of the legal system.

Maverick has the greatest overlap with larrikin, and can be seen as the US equivalent. Zwicky for instance was called a maverick. Maverick is said to come from Sam Maverick (1803-70) an American cattleman whose lifetime spanned the 'civilising' of the American frontier. According to some, he refused to brand his calves, which came to be called mavericks. Both larrikins and mavericks bend rules rather than break laws.

Thus far, we have ranged the terms along a continuum according to one dimension. Other dimensions give a more complex picture. For instance, these terms differ in their relation to equilibrium conditions. Outlaw and pioneer refer to far-from-equilibrium conditions found on frontiers. They are romanticised in a way that is harder with criminals, because of the alternative space they occupy compared to the settled, close-to-equilibrium present (Hobsbawm 1969). In this respect they have something in common with pioneers, explorers and innovators, in new territories before the rule of law has stabilized.

In the following diagram we give a simple map of this field of meanings:

Semantic space of larrikin princple

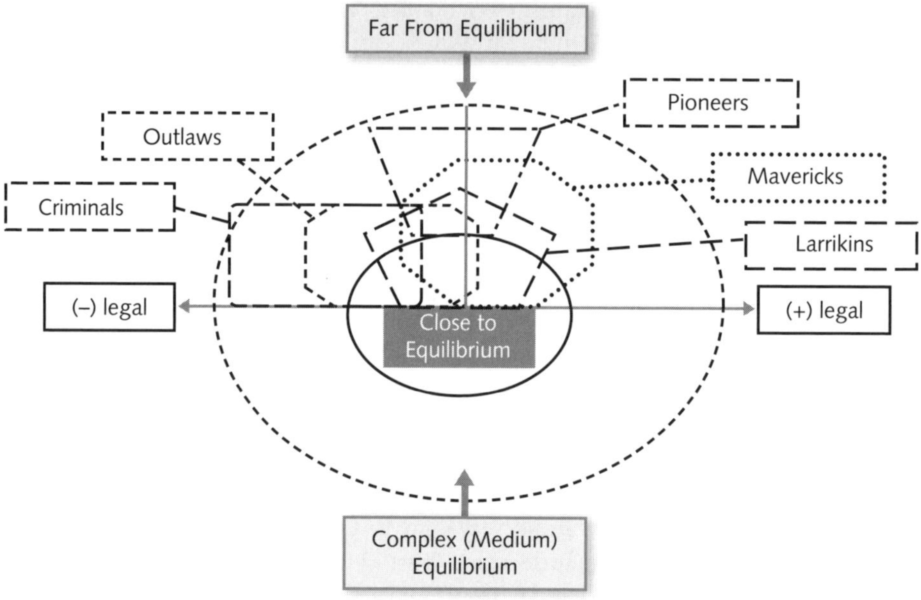

[Diagram: A field showing "Far From Equilibrium" at the top with an arrow pointing down to "Close to Equilibrium" (center oval), and "Complex (Medium) Equilibrium" at the bottom with an arrow pointing up. A horizontal axis runs from "(−) legal" on the left to "(+) legal" on the right. Labels positioned around the field include: "Outlaws" and "Criminals" on the left side; "Pioneers", "Mavericks", and "Larrikins" on the right side. Overlapping dashed shapes indicate fuzzy regions for each category.]

All these terms and qualities overlap with all the others, so that each has the qualities of the others to some degree. Some larrikins are very like some mavericks, sometimes but not always. Larrikins have a dark side, a connection with criminality, though not strongly.

We also represent equilibrium-conditions in simplified terms, as two circles: medium equilibrium and far-from-equilibrium. For instance, outlaws overlap with criminals in illegality, but operate further from equilibrium, where rules and laws are uncertain. Larrikin meanings differ depending on equilibrium conditions. Close-to-equilibrium, they are distinct from criminality, but further from equilibrium there is greater overlap.

iv) Fissured fields. Fields of meaning are not only made up of complex, overlapping fuzzy meanings. Conflict is everywhere in social life, and always affects meaning. Semantic divisions break up fuzzy flows, reflecting conflict and lack of agreement even when they apparently refer to the same thing. We illustrate with an analysis of a common object, the boss/manager, from two points of view, managerial and worker/larrikin perspectives.

To focus our discussion, we use a recent article in *Time*. Jeffrey Kluger reported findings from an academic study (Anderson and Kilduff 2009) that showed 'there's more reason than ever to suspect that your boss is faking it' (2009:35). Kluger claimed this study exposed the flaws of the 'all-hat-nocattle leader'. Often those who presented as leaders were less creative and competent than low-key group members. Kluger concludes:

> None of this may come as a shock. You don't have to be a homeowner burnt by the housing fiasco to know that the people in charge don't always deserve to be. But if you think your boss is making stupid moves, maybe the best thing you can do is say so. Loudly and frequently. The result could be your getting promoted – or fired (2009:3)

Kluger uses larrikin discourse, everyday language lacking respect for bosses. If we compare his text with the original article, we find a surprising lack of semantic overlap. The academics do not talk about 'bosses'. Their key terms are 'dominance' and 'influence'. The semantic field of management textbooks is different again. Their key terms are 'management' and 'leadership'. They sometimes use 'influence', and also 'boss' at times, but not as key terms.

This shows an important fact about language in sites of conflict. Even when people seem to be talking about the same thing, they may use different words in semantic fields so differently that interpretations hardly connect. They will probably act as though they see a different reality.

But if this problem is recognised it can be resolved, if there is a will. The three perspectives can be understood as a three-body system, organised by different relationships to the common object, management. In this three-body process, each perspective affects the other to create a fuzzy, complex meaning. So 'boss' can be understood fuzzily as both manager and not-manager, dominant and not-dominant. As Zadeh insisted, fuzziness exists first in human thought, and only afterwards in language. That is a vital lesson for managers, larrikins and academics alike.

Larrikin myths and meanings

Narrative analysis has become an important strand in organisation studies (e.g. Czarniawska-Joerges 1998, Gabriel 2000). Here we draw on the work of French anthropologist Claude Lévi-Strauss (1963). Lévi-Strauss's theory, combined with chaos theory, suggests productive ways of analysing myths, stories and organisations and the role of the Larrikin Principle.

Lévi-Strauss's key insight was to see myths as logical operations, representing and resolving, in imagination, the most profound and intractable contradictions in the categories of a society, growing out of insoluble problems in social and cultural life.

Although the stories are often untrue in detail, the problems and contradictions themselves are profoundly true. That is why myths play a key role in education in 'primitive' societies. Societies of all kinds, throughout history, may change the contradictions they battle with, but they do not escape the fact of contradiction.

In many examples from so-called 'primitive' cultures Lévi-Strauss showed that apparently strange or bizarre elements in stories made sense as a way of capturing and dealing with these contradictions. British anthropologist Mary Douglas (1970) followed this theme in her analysis of 'dirt', which she defined as 'matter out of place'.

'Dirt' provokes acute ambivalence in every culture, expressed in taboos, objects which are strongly forbidden. Douglas showed that the same thing under some conditions can signify immense power. 'Dirt', like myth, mixes categories that should not be mixed, except by gods and leaders who must understand and manage such power.

Lévi-Strauss saw an especially important role played by anomalous figures, relationships, places and actions. One such hero was the Trickster. There are many tricksters in ancient myths. For instance the Norse trickster god Loki was a fire god whose name is probably derived from *logi*, fire.

In Roman and Greek society, the Trickster figure had links with commerce. Trade happened along the dark margins of society, not incorporated within the dominant political regime. The Celtic god Lugh may have etymological links with the larrikin. Julius Caesar identified Lugh with Roman Mercury, god of commerce, another trickster god. Mercury is usually linked with the Greek Hermes, god of boundaries, and the Egyptian Thoth, god of learning.

None of these gods is identical or stable. To complicate matters, among the Scandinavian gods, Mercury is usually linked not with Loki but Woden, who became a supreme god, more like Jupiter. Within and across mythological systems they form a fuzzy set that changes over time, linked by common characteristics, always with distinctive features. Each plays a part in a different set of stories. All are ambivalent gods, often dangerous, partly outside social norms and constraints, but ingenious and creative, sources of arts and sciences. They are the soil out of which modern trickster figures have grown, including the larrikin.

In myths recorded in historical times, the hero is usually a man. But some of these myths came from early times, when the supreme deity was a woman, the mother goddess (Gimbutas 1991). Egyptian Thoth, for instance, was a moon goddess. Greek Hermes combined with Aphrodite, goddess of love, to form the ambiguously gendered Herm-Aphrodite. This ambiguity is well-suited to the modern age, one of whose characteristics has been the globalisation of feminism. Larrikinas may be older than larrikins.

The scope of the Trickster figure provides a historical basis for extending the Larrikin Principle outside Australia, beyond English-speaking societies. In ancient Hebrew culture, the prophet and the scapegoat played analogous structural roles. In these terms Jesus Christ, prophet and scapegoat, was a larrikin ancestor. The historical Chinese general Guan Yu became fused with similar meanings to become a kind of god of commerce for overseas Chinese

(Hodge and Louie 1998). Hanuman, Hindu monkey god, is another creative figure and trickster god, tamed in Buddhist traditions.

The Trickster god is found in many ancient societies. Australian Aboriginal culture has the *maban*, 'clever man'. In North American Indian traditions, Coyote is ambivalent, cunning but dangerous. In the Maya mythology of Mexico, the Popol Vuh, monkey twins Hun Batz and Hun Chuen played a similar role, patrons of the arts, including dance and writing (Anonymous, 1947). Examples in Brazilian traditions include Kanassa, who like many Trickster figures stole fire from the gods.

Lévi-Strauss found this kind of figure crucial in traditional mythologies to help people to think about the core dilemmas of the society. Their anomalous meanings and mediating functions live on, as necessary now as ever for dilemmas of global capitalism. Laid-back, flexible Larrikin(a)s can move between present and past, across the globe, between rich and poor, powerful and oppressed, as s/he has always done.

Uses of history

These patterns only make sense if we understand how they came to be. History matters. Contradictions in the contemporary Australian larrikin flow from its immediate history. The social reality behind the term came to Australasia after 1788, when the English invaded Australia and claimed it for the British crown, displacing or killing the original inhabitants. Soon afterwards they also invaded New Zealand. This new society at its first moment of existence in Australia was already a three-body system: two extremes of British class society, governors and warders on the one side, convicts and servants on the other, plus Aboriginal inhabitants.

The first 'larrikins' belonged to the second group, infected with rebellious thoughts by political prisoners amongst the convicts, especially from Ireland. But the Australian convict system was meant to be enlightened, aiming to reform convicts as well as punish them. It did not always work out like that, but many convicts won their freedom and succeeded in their harsh new land. Over time, divisions became blurred. Warders became larrikinised, convicts developed an inner warden. In spite of their conscious efforts, both groups affected and were affected by Aboriginal Australia (Hodge and O'Carroll 2006).

Over the 19th century, the two dominant elements evolved together in a single image of national identity which drew on both. The 'larrikin' element shaped identity, while the old colonial ruling class still monopolised power. This dual structure was a cybernetic system. The radical, egalitarian values of the larrikin element counteracted the tendencies of the elite, still formed by their British connections, still bound to the British ruling class. Without the Larrikin loop, Australia could not have been imagined as a cohesive society, separate from Britain.

A version of this history is still alive and well among Australians. Here is Denis:

> I think it was something that was based to an extent on a historian's view of our history or European history in Australia, i.e. that it was certainly in New South Wales, Tasmania, one of the convict origins, where that type of approach, that larrikin approach, was the way that you got things done. It was the way that you survived. You had to break rules, you had to find out ways of doing things because otherwise you could starve and in some other ways be persecuted.

History, even in this brief form, makes sense of some of the twists and turns in the meaning of larrikinism in Australia. For instance, in 1878 Ned Kelly, son of Irish immigrants, formed a small gang of 'outlaws', as they were called.

For two years this gang survived in the bush, evading capture. In one episode they robbed a bank at Jerilderie, where they burnt mortgage bonds held by the bank. In Ned's day, the banks made exorbitant profits through interest rates that were too high to be repaid. Popular resentment against the banks still bubbled up from under the surface in the 2008 crisis, but President Obama's administration did the opposite to Ned. They bailed out the banks instead of robbing them.

In spite of a short, unimpressive career as a criminal, ending on the gallows in 1883, Ned became an iconic figure. He was called a 'larrikin' at the time, when the word meant 'delinquent', without any positive connotations. 100 years later he was a national symbol, celebrated in art, literature and film, where he was portrayed by Mick Jagger in one version and the late Heath Ledger in another.

A thin chorus of voices expresses the old establishment view that Ned was just a criminal, but these voices have no traction in Australia. The transmutation is complete. In 2001 Neo-Liberal Prime Minister Howard, populist Australian rather than Neo-Liberal, exerted his influence to give larrikin Ned a prominent place in Australia's new National Museum. In 100 years, Kelly went from historical criminal to national myth, yet he is still both, to some degree. History always produces fuzzy meanings. *Larrikin*'s meaning is still fuzzy and unstable.

The USA has many features in common with Australia. Both nations are former British colonies, formed by invading and 'settling' territories occupied by native peoples. But Australia got its identity from dissident convicts while US identity comes from the Puritan Fathers, dissidents from Britain, but more like Australia's warders than convicts.

US history is longer, producing more complex patterns. The nation was in a sense re-founded with the expansion west in the mid-19th century. From here emerged the figures of the pioneer, the outlaw and the maverick, terms which come closer to the Australian *larrikin*.

Yet, perhaps, the differences are not so great. Kennedy is the 20th century US President who achieved closest to mythic status. He also had Irish Catholic origins, and a father with dubious, if not criminal, connections. He became the first Catholic to win the Presidency, defeating Nixon, an ex-Quaker. In the 19th century he would not have been able to overcome the Protestant tribalism of the USA, any more than Kelly could have in Australia. A seismic shift has happened in both nations.

In 2008, Barak Obama was elected the first US President with African and White heritages. All commentators hailed this as another and greater seismic shift. The two heritages are separable in his history, black Kenyan father and white US mother, specifically and consciously mediated in his person, not fused, as with most Afro-Americans. The Larrikin Principle suggests that this quality may make him seem especially dangerous, or special.

The colonial factor

The Australian larrikin story unfolded within a larger, longer history of inequality and globalisation. William Easterly was an employee of the World Bank, a leading instrument of Neo-Liberal aid policies. Drawing on decades of experience Easterly (2006) argued that inequalities in the world today derive from the conditions present when they were first founded. These inequalities, he says, actively produce and intensify disadvantages found in the most unequal nations. He contrasts this with the promises of Neo-Liberal economics, that the same policies disseminated globally will benefit all nations equally, slowly minimizing differences between them and within them.

For Easterly the key difference in patterns of colonisation is between 'Settler' nations, where settlers formed the majority of the population, and non-settler nations, where a small elite dominated a larger native population, sending wealth back to the home metropolis. The first kind, he says, produced a larger middle class, which demanded stronger and more open institutions, and more education. The second kind polarised into a small wealthy elite and a large impoverished majority, with only a small middle class. This tendency he finds has been self-perpetuating. The descendant societies of settler societies have far more wealthy middle classes, and a higher GDP.

For all the differences between the USA and Australia, both are settler societies, and both fit Easterly's pattern. This is the background of one larrikin narrative. The other two nations we look at, Brazil and Mexico, fit Easterly's pattern for non-settler nations. In our data we find a different pattern of larrikinism related to this generic difference. In Mexico for instance, initially there seems no obvious equivalent of the Australian larrikin.

Larrikin qualities are distributed between a number of terms and figures, polarised between good and bad.

Yet there are points of contact. We will take the complex case of Marcos, leader of the Zapatistas, an indigenous movement which emerged in 1994 in the jungles of Chiapas, an impoverished region of southern Mexico. The trigger for this uprising, according to Marcos, was the impact of Neo-Liberalism on Mexico's poor, betrayed by Mexican elites who continued a 500-year colonial tradition under the banner of Neo-Liberalism.

Marcos has many larrikin features. One typical comment is his 1997 response to questions about whether he was gay, recorded in Wikipedia. Yes, he said, Marcos is gay:

> Marcos is gay in San Francisco, black in South Africa, Asian in Europe, Chicano in San Isidro, anarchist in Spain, Palestinian in Israel, indigenous in the streets of San Cristóbal... dissident against Neo-Liberalism, writer without books or readers, and for sure, a Zapatista in South-east Mexico.

This is larrikin wit, mocking himself and others, a serious Trickster committed to social justice. He creates global scope for his Marcos persona, weaving different oppressed minorities across the world into a fuzzy, capacious identity.

In Mexico he is called *insurgente* or *revolucionario*, words with a long history of serious opposition, yet he is a revolutionary with a difference, and an exceptionally successful leader. With few material resources, in 1994, he took on the might of the Mexican state and survived. Since then he has become a major presence in Mexico's political landscape, and his critique of Neo-Liberalism is a global export.

Easterly agrees with Marcos that effects of colonialism still shape the Neo-Liberal world. In terms of a framework like this, we can see that many important differences, in the form taken by the Larrikin Principle across the globalised world, reflect systemic differences in that world, more than differences in the principle.

Larrikin politics: Is Rudd a larrikin?

Prime Minister Kevin Rudd is Australian and now opposes Neo-Liberalism, but that does not make him a larrikin. For instance he entered politics with a reputation as a 'Brain' (First class Honours in Chinese, fluent in Mandarin) and an obsessive workaholic: 'Nerd' not larrikin. Yet he has convict ancestors, and was raised a working class Catholic. For image-managers, being a Nerd made him seem unelectable as an Australian political leader, and he cultivated a 'more human' image.

For instance, in a TV interview on March 8 2009, the larrikin showed through. He defended his government's massive injection of bailout funds:

> People are going to run a huge scare campaign about government debt and government borrowing. People have to understand that, because there's going to be the usual political shit storm... Sorry, political storm over that.

The Prime Minister of the nation covered his mouth, as if embarrassed but still enjoying the moment. The informal language broke the political rules of TV interviews with public figures, even in Australia. They connected viscerally with many ordinary Australians, but sometimes negatively.

They carried a risk, even in Australia. Robert Doyle, Lord Mayor of Melbourne, member of the Liberal (Conservative) party, took him to task. 'I think it was a carefully scripted attempt to make himself appear human, one of the lads.' The idea that this larrikin is a fake would be potent criticism by his enemies, if it stuck. On this occasion, the phrase 'one of the lads' is British vernacular not Aussie slang. It puts the speaker on the Establishment (British/Warder) side, not a good place to be politically, in Australia. That is why Doyle attacked the larrikin image in the first place.

Cybernetics of the larrikin

Cybernetics is a helpful way of representing the Larrikin Principle in generic terms. Weiner took the term from ancient Greek *kybernetes* helmsman, guiding boats with subtle, skilful shifts of the rudder. We update the image with modern equivalents. Alongside the *Cybernetes* of the ancient Greeks, we include the surfer, iconic figure in Australian (and US) culture. Australians learned the art of body-surfing from Hawaiians, originators of the modern sport. It was once a male domain, but no longer. Australian Layne Beachley was 2008 women's world champion. We could equally well have included cowboys, who manage horses with the same art. Interestingly, *manager* probably comes from Italian *maneggiare* from *mano*, hand: to guide a horse with subtle use of 'soft' hands. There is cybernetic wisdom in the etymology.

The cybernetes-surfer-larrikin complex system

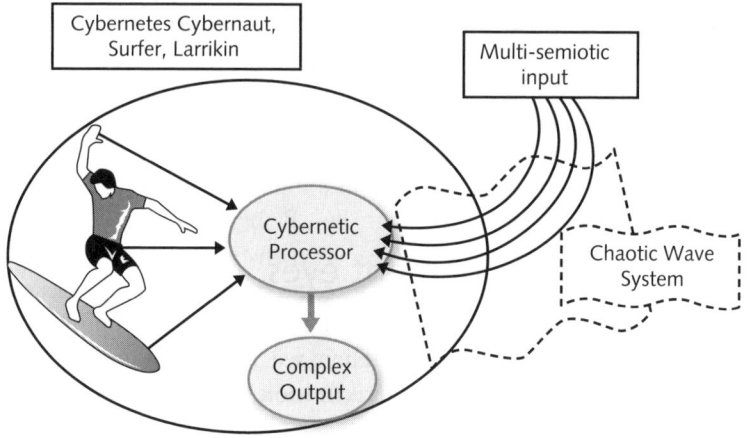

In this picture, expert surfers take account of many subtle signals through many senses, routed through kinaesthetic senses in their body, adjusting the angle of the surfboard to maintain balance and momentum. Those signals come from the surface of the ocean they are in contact with. The energy of the wave produces transient regularities in contact with the changing bottom of the ocean as the surfer system passes over it. Surfers react to small signs from the ocean and the wave, most too small to be detected by anyone but an expert surfer, none recorded in precise, quantitative terms. The energy comes mostly from the wave and the ocean. The surfer's skill captures and directs it, but does not create it.

Zadeh's fuzzy logic was designed to make better control systems for complex and unstable conditions, such as cybernetes in stormy seas, surfers riding big waves, cowboys riding wild horses, or larrikins in modern business organisations. Fuzzy systems work because they incorporate a range of qualities, as in the larrikin. Cybernetics pioneer Ross Ashby formalised this in his Law of Requisite Variety:

> If a system is to be stable, the number of states of its control system must be greater than or equal to the number of states in the system being controlled (1956: 124)

This cybernetic principle explains why the diversity of larrikin attributes is so vital to every organisation, and why good managers need it. Drawing on Lévi-Strauss, we add a further aspect of this diversity, equally essential in myths and organisation: diversity must incorporate the key contradictions.

Ashby's law applies to all kinds of diversity. Gome and Ross (2003:39) criticised the gender balance in Australian organisations compared to the USA. In 2001 women held only 8.4% of board seats. The US figure is

12.5%. Comparable data for Australia, Brazil, Mexico and the USA is hard to find, but the World Bank gender statistics (2009) report women's participation in managerial positions as: USA 46%; Australia 36%; Brazil 31%; Mexico 25%. Brazil is not far behind Australia.

Ashby's law presents this as an issue of cybernetics, not social justice. In terms of the Larrikin Principle it is both: a less good system precisely because it lacks justice.

3. Larrikins in management eyes

It is time to invert our gaze, to see what the Larrikin Principle looks like from a management point of view. Then we invert the gaze again: what does this management gaze on the larrikin look like from a larrikin point of view?

We will use our management textbooks as data again. Here is Davidson's definition of Management, similar to that of the other textbooks:

> **Management** is a set of activities (including planning and decision making, organising, leading, and controlling) directed at an organisation's resources (human, financial, physical and information) with the aim of achieving organisational goals in an efficient and effective manner (2003:5)

Efficient is defined as 'using resources wisely and in a cost-effective way', and **effective management** as 'making the right decisions and implementing them' (p. 6).

This has the crisp, categorising mind of Managerialism (4 types of activity, 4 kinds of resource), emphasising efficiency and effectiveness. It creates a re-assuring impression of order underpinning effective and efficient action, with no need or place for the Larrikin Principle. However, a larrikin might notice the fudged language, the unscientific 'wisely' and 'right'. Who decides what is wise or right? The boss? Oh yeah, the larrikin might say.

As we saw, this concept of 'Management' links all managers with CEOs in the fuzzy space of the management Ideological Complex. Davidson does address this issue. He reproduces a model found in all texts in some form: a pyramid in three layers. 'Top management' is at the top, then 'middle management', a large group 'primarily responsible for implementing the policies and plans developed by top managers'. Below these are 'first-line managers':

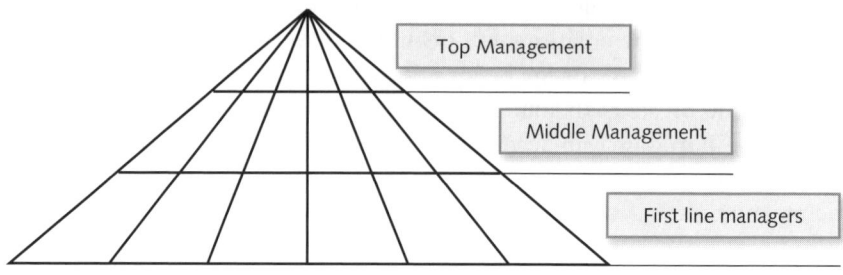

(Adapted from Davidson & Griffin 2003:13)

Davidson makes a subversive larrikin comment on middle management. He notes that in Australia they form 5% of the workforce, but that they have recently (2003) suffered 17% of retrenchments. That makes them look like victims not agents of change. Larrikin readers may note these statistics and begin to think. Maybe obedience to organisational goals will not be rewarded. Maybe they will need larrikin scepticism and flexibility to survive, in their present job or the new one they will have to find.

Other percentages carry messages. For instance, managers form 10% of a workforce. The rest are 'workers'. Yet this image of 'management' does not include these other employees, except in the abstract form of 'human resources'. This image of the object of Management Studies leaves out 90% of most organisations. Coincidentally, this is a similar proportion to dark matter in galaxies, and omitting it has similar effects. It gives prospective managers a distorted image of the organisations they will join.

This pyramid image is so popular and taken-for-granted in management circles that it is worthwhile to unpack it with a larrikin gaze. Pyramids are good shapes for monuments to dead pharaohs, but not such a good design for complex flows. Movement here is in one direction only, downwards and across high boundaries.

Egyptian pyramids rested on ground, as their practical architects knew well. This image does not show the ground which sustains the management pyramid, i.e. everyone else in the organisation. These others are excluded from this image. This makes it a poor image to use for anyone wanting to understand organisations.

Northcote Parkinson provides a different perspective on the pyramid, in his popular satire of bureaucracy. Parkinson was British not Australian, but his style and critique have many larrikin features, including the use of parody and humour, and irreverence towards authority. His 'law' started from the lowest fractal level, a personal observation on a paradox of everyday behaviour, that 'work expands to fill the time available for its completion' (1957:11).

This becomes a paradox of bureaucratic structures; 'there need be little or no relations between the work to be done and the size of the staff to which it may be assigned'. Growth of the pyramid is driven from its middle. Parkinson gives two axioms: '1. An official wants to multiply subordinates, not rivals, and 2. Officials make work for each other' (1957:12).

Parkinson parodies the 'scientific' manner of Managerialism with vacuous mathematical formulae, but he captures the perverse logic of this cybernetic machine. It is designed to produce more work, more prestige and more pay for more officials, rather than do what the organisation exists to do. Generations of managers laugh at the accuracy of Parkinson's satire, yet its larrikin wisdom is not mentioned in any of our textbooks.

The management pyramid also has a context. As Davidson knows, organisations do not exist in isolation in a simple world. He emphasizes its complexity:

> The average executive works 60 hours a week, has enormous demands placed on his or her time, and faces increased complexities posed by globalisation, domestic competition, government regulation and shareholder pressure. The task is further complicated by rapid change, unexpected disruptions, and both minor and major crises (2003:6).

This image of managers bears no resemblance to serene pharaohs on top of pyramids. There is tension in management ideology which is reflected in debates and present in all our management textbooks, between 'management' and 'leadership'. Davidson reveals a typical ambivalence about 'leadership', from a managerialist perspective:

> The paradoxes of leadership: it contributes to good management yet goes beyond it, with elements of intellect, mystery, romance and charisma. The challenge is to apply good leadership to help create good management... Organisations need sound management, yet they look for great leaders. One could be forgiven for feeling bewildered (2003:563–4).

Organisations urgently need dynamic, creative, flexible people at every level, to respond to complexities and crises. That means kinds of larrikins, under other names, but still arousing ambivalence in managers.

Davidson and the others deal to some degree with the need for larrikins. When discussing managers' need to know what is going on in the world outside, he talks of 'boundary spanners' (2003:87). These are junior staff, responsible to first-line management, who go into the world outside the organisation and report back. But this is larrikinism on a tight leash. These low-paid, low-status larrikins are easily ignored. Real larrikins would span more boundaries and be less conformist.

The pyramid image is a simplistic, unrealistic, misleading image of organisations. The absence of a place for the larrikin is symptomatic of a profound problem in this management ideology. The pyramid image represents immobility and rigidity, a model based on authority and control.

These would be fatal qualities in any organisation, but this is not how the image is meant to be read. It is a fantasy about how would-be managers would like the world to be: serene, harmonious, stable, symmetrical, with power flowing without effort or resistance. All those inconvenient humans who make managing so complicated simply do not exist.

Whistleblowers

There is a third trace of larrikins in management textbooks. All textbooks now mention whistleblowers. Typically the discussion is brief, but it is there. Normally it is not connected to larrikin features like informal practices. It is an example, often the only example, of the fact and impact of corruption. Whistleblowers are important because corruption is part of the dark matter of management and organisations. We are especially interested in the theme because this is where the Larrikin Principle is expressed in pathological, far-from-equilibrium conditions, almost denied by management theories as well as by managerial practice.

We focus our discussion using a film, a Hollywood product which captures many of our themes, including the fact that, as a popular film, it is larrikin data. The Insider (1999, Director Michael Mann) dramatised a real life story; Dr Jeffrey Wigand's exposé of dubious practices in the tobacco industry. Wigand, as a senior executive at Brown & Williamson (B&W), was the most senior 'whistleblower' to come forward with insider evidence in this industry.

The company brought massive pressure and influence to bear in order to silence him, but finally failed to do so. The film frames his story with another, about a CBS-TV program produced by reporter Lowell Bergman which broke the story. The tobacco giant was able to block the program for a while, before its efforts were defeated. Outside the film, the tobacco giants lost an even bigger battle. It had to pay USD 368 billion to settle a case brought by the State department and other agencies.

The historical Wigand played a significant part in this second battle, too, providing the prosecution with continuous high-level advice and analysis. So valuable was his input that the prosecution insisted, as part of the settlement, that B&W withdraw all cases against him.

The film dealt with two partial larrikins in two spheres of life. Wigand was powerful and dangerous because he was a hybrid: a very senior, capable manager (USD 300k p.a. with perks) and an ethical scientist. The film named him an insider rather than a whistleblower, which best captured the ambiguity of his status: deep inside the industry yet also on the outside. He said, in a statement to a working party on whistleblowers (Wigand 2007), that he took the B&W position for the salary and because he wanted to use his research expertise to produce safer cigarettes.

This win-win aim is central to the management Ideological Complex. It turned out not to be what happened in practice. It also positioned him as a Lévi-Straussian hero, dangerously anomalous, a Trickster.

The Ideological Complex of the mass media includes both profitability and journalistic ethics, but again the reality proved different for Burgmann. The owners of CBS included tobacco executives who used all the power at their disposal to kill the first story. He had to rebel against his bosses to get the story on air, and to use a range of tricks, some dirty. He too was an idealistic Trickster.

Yet there was a positive outcome. This David-versus-Goliath victory, against all odds, was a butterfly-effect. Butterfly-effects are unpredictable but need a far-from-equilibrium context to happen. In this case, the larrikins connected two bodies of a three-body system. The third was the state. All three needed to come together for this outcome. The film mainly represented the action between the worlds of business and the media, but Wigand's capacity to work closely with government was also decisive.

In evidence to the House committee Wigand emphasised networks as crucial:

> I want to make very clear that I am able to be here today – not because I was protected by any whistle-blowing statute – but because of the courage of so many people (2007:1)

This side of the story is not commented on in the film, yet a crucial part of the Larrikin Principle is relations of trust and support. Likewise the line Bergmann refused to cross was his absolute commitment to his sources. For both, the issue of trust was what drove them to bend or break rules.

As a film, The Insider told only part of the story. It also used the resources of film as a medium to add layers of meaning. One layer is casting. Al Pacino was cast early on for the film as Bergmann. As a well-known actor he brought expectations to the role. He made his mark playing mafia figures, criminals. In the role of an idealistic reporter this persona gave an edge to the meaning of the character. Bergmann in the film was, fuzzily, both a reporter of integrity and a darker, more dangerous force.

The casting of Russell Crowe as Wigand was more surprising. Crowe was 33 years old, playing a man in his mid-fifties. He was an Australian/New Zealand type larrikin, who then and later received unfavourable publicity over his larrikin ways. Yet his performance as Wigand won an Oscar nomination and other major awards. Later he won an Oscar in the Gladiator, as a rebel/larrikin in a Roman tunic. He almost won another Oscar for his role as psychotic Nobel Prize-winning economist John Nash. A larrikin subtext can add value.

Crowe playing this character added contradictions to the film's meanings. In addition to Wigand's mix, insider/outsider, executive/scientist, Crowe's performance suggested a larrikin element that others had not mentioned in Wigand. The filmmaker managed this subtly and skilfully, shooting Wigand in early sequences from the side, in partial frame, alienated and disconnected, lacking energy. In a turning point, Wigand's interview with Sandefuer, the CEO of B&W, he became angry for the first time, and swore. In Wigand it seemed a return to a truer self that had been buried by compromises and power. The worm turned, and revealed himself as a larrikin.

The Larrikin Principle

We have still not pinned the Larrikin Principle using a single, crisp definition. It is more productive as a space in which to explore connections and disconnections of different kinds, at different levels. At its most general, it carries basic lessons from chaos and complexity. Linear systems of control produce chaos by trying to deny it. The Larrikin Principle works with chaos. Every complex system needs something which disrupts and complicates it. The Larrikin Principle incorporates anomalies and feeds them back to make systems more complex and sustainable. Linear systems of control neglect the need for soft forces to hold things together. Social justice as dark matter permeates the Larrikin Principle.

Yet the Larrikin Principle never exists in this pure form. It is always incarnated in particular places, times and cultures, shaped by their needs and characteristics, solving some problems while ignoring or compounding others. We want the Larrikin Principle in its 21st century version, a product of the dominant system it resists while drawing on cultural traditions older than history. That version does not already exist, a template waiting to be applied to fix all problems. We are still discovering it. When we do, we will still find it is not a perfect fit. Then we will use a larrikin phrase: 'She'll be right'.

Chapter 3

Managerialism

Thus far we have talked about 'Managerialism' without defining it. This chapter will not fully define it either, or at least not in any crisp, linear sense. The crucial problem is its complex, unstable relationship to 'management'. Managerialists assume the two are now effectively the same: that Managerialism is the defining form of management today. Any other form of management is just bad management.

Managerialism is such a shifting term that we cannot separate it completely from management practices, good or bad. We called Managerialism an ideological complex, a set of systematic contradictions. Yet this ideological complex corresponds to a set of management practices built around similar but not identical contradictions. The ideological complex and related practices form a functional whole.

Being this far from simple or single object is then made more difficult to grasp because of an added layer. Managerialism presents itself as 'new', as opposed to a set of practices labelled as 'old'. This label declares that what was done in the bad old days is now obsolete, not done by competent managers in successful organisations.

Managerialism claims this 'new' is different in two major ways. It addresses the challenge of ethics, so that modern management is thoroughly ethical and socially responsible. It has renounced the rigid, linear practices of the past in favour of complex, dynamic styles of management.

We agree with some of these claims, to a degree. Clarke and Clegg (1998) refer to a 'paradigm shift'. They and others see a new wave in business theory and practice, corresponding in some ways to that which we claim for chaos perspectives and the Larrikin Principle.

Clarke and Clegg look critically at this change. They quote a study (Pascale 1990) which graphs an exponential rise in the number and popularity of management theories between 1960 and 1990. They see this as evidence for an explosion of new fads, none solid enough to be called a paradigm (1998:18). Unlike others on this theme, they use 'paradigm' critically, distinguishing rhetorical hype from how US philosopher of science T S Kuhn used it in his theory of scientific revolutions (1962).

We agree with Clarke and Clegg's caution. Yet we note that Pascale's curve follows a Power Law curve, a sign of far-from-equilibrium conditions. We argue that a paradigm shift is indeed underway, driving but not captured in these theories. To understand it better we begin outside of management and Managerialism, with a larrikin perspective.

1. Larrikin perspectives on management

We start our exploration again at the lowest fractal level, with a single case, from our own university, though not from our team. For 14 years Dr Zoë Sofoulis was one of the most creative academics at the University. She was also an exemplary larrikina, cross-cultural (Greek-Australian), and cross-disciplinary. She was an innovator in a difficult, and complex area of high priority to the university, connecting academic knowledge in the humanities with important, complex issues in industry.

For instance, she helped pioneer a new way of managing road traffic issues, in collaboration with Australia's largest motoring organisation, NRMA. She developed original methods of investigating cultural aspects of domestic water use. Her interdisciplinary approach synthesized humanities and social sciences with technology studies. She led initiatives and collaborated with a demanding range of industry and academic partners.

In 2007, this highly productive academic resigned from the university. She gave a farewell speech for the occasion, which she has kindly allowed us to use. In it she took the opportunity to reflect on Managerialism as she had experienced it during her time at the university:

> My time at UWS ... coincided with the entire Australian Dark Ages. Causes of regret and exhaustion with the job mainly relate to that era's failures to value and adequately fund higher education, and effects of UWS's various attempts to respond to these failures.
>
> Even before I got here the mutagenic retrovirus first known as TQM (total quality management) had, along with computers, invaded universities, morphing them from places where groups of professionals and experts managed each other collegially, supported by secretarial and administrative staff, to become places where academics did almost all of our own secretarial work, and were turned into 'human resources' reportable to whole new tiers of administration. While I agree it was good for academics to become more accountable, the TQM retrovirus and its mutant offspring can take over the institution's structures to proliferate themselves.
>
> Weakened by the management retrovirus, and compromised by ceaseless political intervention and hostile media, UWS and other universities succumbed to chronic restructuritis. As with MS, symptoms of this institutional pathology can vary from day to day (or restructure to restructure) and can affect institutions at different parts and levels... Showing that restructuritis can occasionally produce positive effects, we had some degree of autonomy in defining the school's project and academic scope, the courses we taught and how they fitted in with courses and subjects in other areas.

She is highly critical of two iconic aspects of modern management practice (TQM and restructuring). Her criticisms are not ideological, but come from experience of management and its consequences. Zoë was a competent, committed first-line manager, head of a new school. She endorses the management aim of accountability, and she acknowledges positives from restructures she has gone through. She is not hostile to the institution she works in, but the contrary. Larrikins are loyal to their institutions. That is why they criticise how they are run.

As a larrikina Zoë writes in an easy informal style, but the analogies she draws with science show a good grasp of current science. In contrast, Managerialism aspires to what it understands to be scientific (Bennis and O'Toole 2005), but usually with no actual scientific content. Zoë understates the extent of her grasp of science, in the larrikin way.

We will follow Zoë's analogy in more detail. The existence of retroviruses was a revolutionary concept, only accepted into science in 1970, following work by Temin and Baltimore which won them the Nobel Prize in 1975. It was resisted until then because it challenged the 'central dogma' of genetics, the principle that there was a strict linear, one-way sequence, running from DNA (the code) through RNA (the messenger) to the manufacture of the relevant proteins.

Temin and Baltimore showed that this process can happen in reverse. RNA in viruses replicates via a DNA intermediary, and is inserted into the DNA of the host. HIV is a retrovirus, so no scientist today would dispute their importance. Zoë unobtrusively shows her knowledge of the science by using a less fashionable example, MS (Multiple Sclerosis).

As a model applied to management theories and practice, this one inverts the usual taken-for-granted sequence from goals and plans (DNA) to means (RNA). Instead, the means produce new ends or goals. This makes it a useful model for thinking about the instrumentalism endemic in Managerialism, which focuses on means, techniques and instruments rather than on the ends served.

In management discourse, goals exist, but are seen mainly as the preserve of the CEO. But what if, as Zoë's metaphor suggests, these means and instruments do not fulfil the aims of the organisation or society, and instead replace those aims with others built into the new means? And what if those means come from a hostile, alien organism, as Zoë thinks the managerial discourses of business corporations are, for the purposes of a university?

She fears another quality in retroviruses, their capacity to replicate throughout an organisation (and sector – she sees the whole Australian university sector as 'infected'). Such replication is a massive, systemic form of change. There is also the problem of mutation, producing even greater change. Retroviruses mutate much more rapidly because they lack checking processes which reduce 'errors' in DNA replication. This makes them harder to attack, since treatment is hard to focus.

Thus far this is a metaphor, not data or theory. Yet it provides a line of investigation that we will follow up in the rest of the chapter. Are TQM (and similar management tools it mutates into) pathogenic agents in organisations? Are restructures as pernicious as Zoë reports?

Larrikin diagnoses of Managerialism

John Quiggin is a professor of economics at Queensland University. He is also a larrikin. The first testimonial on his blog quotes Senator Richard Alston, Minister for Communication in the former Conservative Australian Government: 'I don't know how he came to be a professor, but anyway he purports to be an economist'. This is larrikin inverse pride at being criticised by an authority figure. This larrikin defined Managerialism in simple, non-technical terms in one blog:

> The standard assumption, backed up by the existence of university departments of management and business schools generating 1000s of MBAs every year, is that management is a science on a par with physics, or at least economics...
>
> The central doctrine of Managerialism is that the differences between such organisations as, for example, a university and a motor vehicle company, are less important than the similarities, and that the performance of all organisations can be optimised by the application of generic management skills and theory...
>
> The main features of managerial policy are incessant organisational restructuring, sharpening of incentives, and expansion in the number, power and remuneration of senior managers, with a corresponding downgrading of the role of skilled workers, particularly of professionals (Quiggin 2003).

Quiggin's demolition of Managerialism is delivered in the characteristic dry tone of a larrikin. His first target is the pretensions to science. His criticism is conveyed only through a slight exaggeration ('on a par with physics'), coupled with a joke at the expense of his own profession, economics. He then summarises its central 'doctrine', its claim to be a generic theory with universal scope. As we saw, all our managerialist textbooks say similar things. On these two premises, only his ironic tone is different.

The three 'features' are another matter. These are 'policy' and practice, what Zoë might call symptoms, but they do not follow logically from the principles. On the contrary, they are often ridiculous and dysfunctional. Restructures and bureaucracy are far from central to Managerialism. Sharpened incentives are important, but understood differently. Quiggin does not emphasise contradiction but shows it.

We will examine and illustrate these principles.

i) Restructuring

His first feature is 'incessant organisational restructuring'. Zoë calls it 'restructuritis'. For both these larrikins this is a defining feature of Managerialism, yet it is not prominent in managerialist writing. Of our 12 textbooks, only one mentions it. Symptomatically, Hubbard (2004) lists 'restructure' in his index, but refers readers to 'downsizing'. What he says about it in his text gives us a clue as to why this term is not central to Managerialism. Discussing problems in organisations in declining industries, he notes: 'For these organisations, *downsizing* (often called 'restructuring') or *withdrawal* may be the most appropriate option.' (2004:141).

'Restructuring' has been used so often as a euphemism for cost-cutting that management textbooks may feel they can no longer use it in its normal sense. The same is true of a term even more central to Managerialism, 'rationality'. Like all the other textbooks, Hubbard discusses 'rationality' in the context of 'rational decision-making' (2004:168-77). He does not mention 'rationalise', another euphemism for 'downsizing/restructuring', so wide-spread that it became the defining feature of Neo-Liberalism in Australia, called 'Economic Rationalism' by Australian sociologist Michael Pusey (1991).

Our data was full of examples of this practice. For instance Frank is a middle manager in a firm involved in a takeover. He benefited personally, but still retains a skeptical, larrikin perspective on how and why things happened:

> The same thing with the cost cutting and those kind of things. Like, they will just ask questions, questions, questions. Justifications. And then they will make such ridiculous rules or whatever… Actually they just want to have cost cutting, expense reduction or whatever. It's not about being more efficient, about logically efficient.

ii) Sharpening incentives

Unlike 'restructuring', the idea of rewards and incentives plays a key role in Managerialism. However, it is typically understood in terms of a defective cybernetic model with linear assumptions, in terms of which the only effects of a given incentive are those consciously fed into the design. Tony, a bank employee, describes the more complex reality that can turn linear structures of incentives into unpredictable, chaotic systems. He was on the receiving end of an incentives scheme with a built-in contradiction. In this scheme staff were rewarded in terms of their scores under two columns, one for sales and the other for 'behaviors'. As Tony explained:

> Even if you got all the sales you were meant to get, but if you didn't do it the way they wanted you to… it's a statistical thing, you couldn't fudge that, you got the figures there, you either get them or you don't, but the behaviours (are) purely subjective, according to what your manager thinks.
>
> So if you're getting the results and your manager says "You're not really doing it the right way" or "He wasn't trying hard enough, he was just getting lucky", whatever, then Bingo! You don't get a bonus.

> Because they go, yeah, you got a 4 on your results but you only got a 2 on your behaviours, so that equates to a 3 and that means you don't get a bonus. So that was their way of getting around paying people bonuses at the end of the day.

This bonus system was designed to promote sales, but another priority, cutting costs, cancelled that, because managers were rewarded for that. There might have been higher profits if staff had responded to the incentives with increased sales, but the incentive scheme was too precisely targeted to achieve that. As a result it produced the opposite: an initial spike in results followed by a drop, exacerbated by acute systemic problems of morale.

Paradoxically, larrikin Tony would have preferred the clarity of old-paradigm management to a new-paradigm scheme which seemed sensitive to complexity, taking into account more than just bare sales figures. But linear thinking in line managers turned this 'sensitivity' into a way to avoid paying bonuses. The bank saved money by not meeting its promises; but, after a while, it probably lost money from demoralized staff, who will not trust the next 'bonus' scheme.

iii) Managerial expansion

Quiggin's idea that Managerialism expands the number and scale of managers was already stated definitively by larrikin theorist Parkinson in his 'Law'. One interviewee, Mark, was a manager in a New South Wales colliery which was taken over by a multinational. He had a larrikin perspective on the style of the new owners:

> When it started up it was run basically like an Australian company, and with Australian people running it. And then it was taken over by an American company and they put in Managers, they put in a Manager for everything, you couldn't make it for morning tea. And within a couple of years they folded because there was that much management and bureaucracy, nothing was happening. So it got sold again, changed hands, and it's now in a joint venture type thing, with basically Australians running it. It's making a *motza* ('lots of money').

Mark is not entirely impartial. He clearly prefers Australian to American management styles. But there is no reason to doubt the basic story. The company was taken over, and it went broke and was sold on. Under new (less-managerialist) management it is making more profits. The managerialist style reported here, with many expensive layers of bureaucracy, is not what managerialist hand-books recommend; however, it is an unforeseen but inherent consequence of the practices which grow out of the ideology's obsession with management control.

Frank benefited from his takeover, but had reservations:

So I'm reporting to the person actually very powerful. And there are three of us reporting to that person. No project yet. And the guy I'm reporting to, actually he doesn't really have any idea what we are doing. This is politics.

We came to him, we ask him, what should we do now? He doesn't know. So we ask him what we should do. And he said he doesn't know, and he asked us to propose (something)… But because we are too high in terms of our position, we are too close to high management. Yeah. The politics that we have to play…

We ask him and he throws back the ball again to us. And then we said, okay we will have a proposal. For integrating this and so on. He said okay, make it formal, propose this. And so we went to management…

And in the end, when they approve, they only approve us to continue planning. Our perception, it is also their politics they are playing with us, because they don't have anything for us. They want us to look busy. And they cannot say we don't know what you are doing, and we don't have anything for you. They are also playing their games. Yeah. And with us.

Frank tells us with larrikin candor what is going on, this highly paid work whose only product is the image and standing of one member of senior management. The word from the managerial lexicon, 'planning', has a potent effect even though it is used here to conceal the absence of planning.

Zoë drew on critical management theorist Braverman (1974) to comment on our analysis of the dynamics of Managerialism:

> The idea is that for scientific management to work, the bosses have to rely on the workers providing info and analyses of how the work processes happen. So the managers' capacities to manage are directly parasitic upon the workers' knowledge. This was the dynamic I was emphasising in the restructuritis analysis: how every restructure relied upon the workers to figure out the fine details and necessary workarounds to implement it.
>
> Meanwhile (this was not quite in the speech) the TQM virus had so thoroughly infected the institution and workers we had been turned into more or less self-governing, self-administering, self-monitoring and self-reporting units. I don't know what to make of this, and whether to understand as contradictory or complementary these two tendencies – one towards worker self-management, and one towards top-down line management.
>
> Anyway the key point that I think is left out in your account is Braverman's point about management relying on workers' knowledge to set up management and monitoring processes.

Zoë's larrikin comment reminds us of the contradictions in Managerialism. The (unacknowledged) managerialist intent on deskilling workers to balance the expansion of managers may also have the opposite tendency, to build new levels and kinds of skill in those who are left in order to make the system actually run. But then it does not recognise or reward these essential contributions.

2. The double face of Managerialism

We return to the managerialist account of management in its own terms, to put the two versions together in a single account. We begin by analysing one book, *Manager's Handbook* (Heller, 2002). This is a popular text, which we bought in an airport bookshop. Its dust-jacket announces it is 'both an introduction for new managers and an up-to-date reference for experienced managers and those working in small businesses.'

On first impression, this book seems to demonstrate the triumph of the New Paradigm described by Clarke and Clegg. It begins with the complex individual reader and her (the picture shows an attractive young woman) need for self-development in a complex world. The next section, on organisations, begins with a New Paradigm chapter on 'Learning Organisations' by Bob Garratt, who pioneered this concept.

Garratt's fine chapter has an implicit chaos perspective. He rejects 'traditional' forms of organisations with 'simple, rigid pyramid structures, with well-defined boundaries between "inside" and "outside" worlds.' He is adamant: 'Such thinking is now redundant in today's world' (2003:36).

As standard bearer for the New Paradigm Garratt attacks the old pyramid image, using the same arguments as we did in our previous chapter. 'In a complex and turbulent external environment, such internally orientated organisations will not survive' (2003:43). His solution is to invert the pyramid. This model 'radically changes business priorities', he claims. 'The mind-sets of all directors and managers must change'.

This inversion is literally a revolution. The top becomes the bottom, and the bottom becomes the top. Garratt says that there are good reasons as to why this should be so, and he commands all directors and managers to change their mindset.

This is a radical image of how the world ought to be, but how is it inserted into managerial practices? This is a problem of cybernetics. Does this model influence organisations to become more like this? Or does it help managers to talk the talk but not walk it? If so, it might allow them to resist change because they can now talk so well about it. If this happens (as it may, in spite of Garratt's best intentions) then his image of an inverted pyramid may fit Marx's famous definition of ideology:

> 'In all ideology men and their relations appear upside down, as in a *camera obscura*' (Marx and Engels 1976:42).

Clarke and Clegg assemble evidence to show that business today faces a fundamentally new situation, driven by globalisation and the information revolution. There is a new scale of chaotic processes, they say, and new levels of awareness of it. We agree. But the crucial question remains: how far has the 'old paradigm' withered away? Or does the Managerial Complex contain both New and Old, in an ambiguous, unstable but functional unity?

Total Quality Management as paradigm shift?

This is no easy matter to decide, since it depends on what counts as 'New' and 'Old'. The *Handbook* structure replicates the form of the managerialist Ideological Complex, distributed unequally between New and Old. It begins with two soft complexity chapters, one on 'Developing yourself' and the other Bob Garratt's 'Learning organisations'. But then linearity takes over, with 3 chapters on the real business of organisations, and the attributes that managers need to deal with them.

'Planning' dominates these chapters, on 'Financial management', 'Managing resources', and 'Operations management'. The second, by Colin Barrow, has Total Quality Management at its centre, as an exemplar of modern planning. But does this signify a paradigm shift? Is TQM standard bearer of the New, or carrier of the Old? In this section we will use TQM to show how hard it is to decide.

Clarke and Clegg identify the 'old paradigm' with Taylorism/Fordism. At the heart of the 'new paradigm' replacing it they cite the 'Quality Paradigm', exemplified by the success of Total Quality Management (1998:254).

But Zoë, our resident larrikina, identified TQM as the primary vector of the new retrovirus. Although Clarke and Clegg call it an example of the new paradigm, they also report a judgement (1998:251) that TQM is just 'Flexible Neo-Fordism' (Neo-Taylorism). Clearly we need to define the terms fuzzily: TQM is both new, to a degree, and old, also to a degree.

Seeing this key term as fuzzy opens up a series of key questions about the status of the two paradigms. We illustrate by following up TQM in other sections and authors in the *Handbook*. In his general introduction, editor Robert Heller is eloquent:

> Total Quality Management, for instance, enshrines basic truths: that each and every practice and process can be significantly improved for ever; that this improvement can only be achieved if before-and-after results are meticulously measured; that planning and implementation should be entrusted to those directly involved; and finally, that the involvement and the improvement have an energising effect on the whole enterprise – and a wonderful impact on both the bottom line and on the top. (2003:10).

We larrikins cannot control our scepticism when we hear appeals to 'basic truths', especially applied to a tradition less than 30 years old. A word like 'enshrines' increases our suspicions. Is this a new religion?

The list of 'basic truths' does not ease our concerns. The idea of 'continuous improvement' does indeed come from religious beliefs, the Protestantism that shaped Western forms of capitalism, according to Weber (1948). As an idea it comes up against warnings from cybernetics and chaos theory. TQM is a cybernetic system composed of cybernetic systems. Any tendency continued to infinity is likely to be captured by a power law. Economics has known the Law of Diminishing Returns since David Ricardo's classic work of 1817 (2007), which will slow down any such linear progress. Heller's 'basic law' just re-states the old fantasy of perpetual motion.

The faith in meticulous measurement is also Old Paradigm thinking, which comes up against Zadeh's Incompatibility Principle. This warns that precise measurement becomes meaningless or irrelevant in far-from-equilibrium conditions, characteristic of the New. In this aspect TQM is only a new way of ignoring the truths of the New.

'Managing resources' by Colin Barrow is built around TQM models, embedded in linear ideas of planning. He begins by emphasising linear analysis, 'dividing the task into distinct components that can be handled separately by specified teams or individuals' (2003:83). This comes straight out of Taylorism. Everything must be broken down into components and precisely specified: 'An action plan is needed for each component of the overall plan, and for each individual involved in the various teams' (2003:84).

The science part of 'scientific management' comes in, but ambivalently: 'Planning should be done with a scientific rigour, even though it is not rigorously scientific. However, the results should always be presented scientifically' (2003:83). Planning it seems should be only fuzzily scientific: scientific but not too scientific, or more exactly, pretending to be scientific without being so.

TQM is then offered as a model for planning. It 'embodies all the essentials of planning' (2003:86). TQM has a New Paradigm concern for soft qualities like 'employee satisfaction', but this and all other qualities do not exist unless they can be measured: 'TQM also involves seeking metrics for all other activities, since what goes unmeasured cannot be demonstrably improved' (2003:86). At the heart of TQM analysis as recommended here is a rigorously linear form of analysis.

For instance, there is a 6-step analysis, supported by a graph with 6 boxes linked by arrows in one direction. A basic process in TQM is the 'PDCA' cycle: Plan-Do-Check-Act (2003:88). This is modelled as a circle, but is really a linear sequence, a line that returns to the beginning where it can start the sequence again.

TQM was evaluated by two Australian researchers, Samson and Terziovski (1999). They found it to be successful overall in improving performance. However, most of that improvement came from 'soft' factors such as staff morale, rather than from the highly linear 'tools'. That suggests that 'New Paradigm' components in TQM are masking and legitimating the continuing 'Old Paradigm' elements embedded in it.

Our analysis of the *Handbook* suggests that the 'paradigm shift' has not yet happened in mainstream management. The Old Paradigm (linear thinking, exclusive focus on efficiency) is still dominant, though accompanied by some elements of the New (social awareness, chaos perspective). It pretends to be more New than it is.

Yet this book does not present this ambiguity and associated contradictions as problems. Described like this we might suppose that Management theory is confusing and schizogenic. But there are no signs that these contradictions are seen in this way, or seen at all. It is more likely that the shifting, contradictory status of new and old in this up-to-date seamless package shows how smoothly the managerialist ideological complex works in normal practice.

Taylorism revisited

If the 'New Paradigm' turns out to be not so new, we will need to look again at the Old, to see what it is and if it has indeed been buried. The name most associated with the old is Frederick Winslow Taylor. Born in the USA in 1856, he wrote a slim book in 1911 which became the manifesto of what was immediately seen as a new form of management. The style of management he proposed has been called 'Taylorism' after him, no small honour. Its principles have much in common with those who drove Henry Ford to his success at the same time. 'Fordism' is roughly equivalent to 'Taylorism'.

Writers often refer to the present era as 'Post-Fordist' (e.g. Harvey 1989). They could just as well call it 'post-Taylorist'. Either term exaggerates the degree to which management has moved beyond this model, except at the edges. That brings us back to Taylor, with a new need to understand what he stood for.

Why does management think that it is scientific? Taylor is a key source of the unfounded but taken-for-granted role of science in management. Yet nowhere in his classic work, entitled 'Scientific Principles of Management,' does he define science. He wrote when Einstein was publishing his revolutionary ideas, when Marie Curie was pioneering the study of radiation (she won her second Nobel Prize in 1911, the same year as Taylor's book) but Taylor mentions neither.

His 'science' was a marketing tool, a brand name, not an intellectual discipline. The triumph of this marketing strategy relied on its capacity to become a dominant discourse (discursive regime in Foucault's sense 1971). Even today, managers suppose that he put the 'science' into 'scientific management', where it has stayed ever since.

In his major work, Taylor paradoxically writes more like a larrikin than a scientist. He has an easy, informal style, with no graphs or pie-charts, and only one formula. He grounds his argument on stories and anecdotes. He even includes an autobiography. In it he claims that his ideas were born in

1878, when a depression forced him to work as a labourer in a steel factory. This is a classic myth of the hero, a boundary-rider between two groups or cultures, a mediator. And also, myth analysis warns us, a Trickster.

In the same year that he descended into his underworld, as a labourer, Australian larrikin icon Ned Kelly was driven into the bush, to become an outlaw and a legend. In the same year that Taylor became manager, Ned Kelly was hanged. The two paths diverged as far apart as two paths can, yet there is a meaning to the inverted symmetry. Both were kinds of revolutionary, in opposite directions.

Using Zoë's image of Managerialism as a retrovirus, a case could be made that Taylorism was the original form, with TQM and later forms being mutations. In dismantling the Taylorist retrovirus we will begin with the logic of his exposition, to identify and isolate the dangerous mutation. He starts with basic premises. These make no mention of science:

> The principal object of management should be to secure the maximum prosperity for the employer, coupled with the maximum prosperity for each employé. (1967:9)

We notice that he uses the pretentious French 'employé' rather than vernacular American. Equally symptomatic, he stops using it after the next page. From then on they are just 'workmen'. 'Management' here is different from both employers and employees. It is a third force. His formula offers what is now called a 'win-win' solution to conflicts between capital and labour.

This commitment to workplace harmony comes at the beginning of his theory, before science has been mentioned. It is not an after-thought. Taylorist management must produce harmony as well as efficiency, social as well as material good.

Taylorism and the managerial ideological complex

Yet, Taylor's system unfolds in another direction as he develops it. In a famous statement he elaborated on his view of workers, which gave him reason and justification for his system:

> Now one of the very first requirements for a man who is fit to hand pig iron as a regular occupation is that he shall be so stupid and so phlegmatic that he more nearly resembles in his mental make-up the ox rather than any other type. The man who is mentally alert and intelligent is for this very reason entirely unsuited to what would, for him, be the grinding monotony of work of this character. Therefore the workman who is best suited to handling pig iron is unable to understand the real science of doing this class of work. He is so stupid that the word "percentage" has no meaning to him, and he must consequently be trained by a man more intelligent than himself into the habit of working in accordance with the laws of this science before he can be successful. (1967:59)

Although he talks of other kinds of work requiring other kinds of ability, this is the example he begins with, a working man whose broad shoulders (and almost non-existent mind) best support the need for scientific management.

It constructs a sharp, binary opposition between labour and management. The Scientific Method which these workmen most need, by definition, they are unable to appreciate in practice. In order for a 'win-win' outcome, they must be coerced or tricked by management.

This contradiction is basic in the theory. It has the same essential form as the managerialist ideological complex. The rationality of scientific practice, faced by the inherent limitations of workers, must be translated into irrationality. The scientific method must be accompanied by another package, a set of tricks to conceal from them what is going on. All this is ostensibly in the name of the mutual benefit to management and workers that scientific management entails.

There is no feedback built in to ensure that it delivers sufficiently equal benefits. Workers must trust the good will of managers behind their manipulative rhetoric. There is enough larrikin in most workers to reply: Not bloody likely!

Science is the main attribute of Taylorist managers which justifies their right to rule. This passage emphasises this right over workers, but it also applies upwards, to employers and owners who may not appreciate the science either.

Taylorism provoked riots when it was first introduced, from workers who did not appreciate their gains from the win-win. It also provoked resentment from some bosses. Taylor himself parted company from his own employers, Bethlehem Steel, because they were not convinced of the merits of his method.

The idea of science that served this rhetorical purpose best is a stripped-down version, uncomplicated by reference to any scientific practice or content, past or present. Instead of science as a vast, complex body of knowledge (including the work of Curie and Einstein), it is reduced to a single method.

Taylor as butterfly effect

Frederick Taylor was remarkably influential. To no small degree he created Managerialism and gave it an aura of scientific status, yet he had no academic base or senior management position. This makes him a good example of Edward Lorenz's butterfly effect (1993), showing extreme 'sensitivity to initial conditions'. In this kind of analysis we ask: why Taylor and not others? What was it in Taylor's background that led to a theory that was to have such extraordinary effect?

This question pushes us to the disciplines that try to explain events at this fractal level (personal biography, filtered through psychological analysis). Laura Calderon, one of our larrikina research assistants, includes therapeutic expertise in her skill set. She offered this psychological analysis of Taylor:

> I thought it might be relevant to point out: how his doctrine seems to be a reaction to a shaming experience of loss of social standing.
>
> It seems to me that he was unable to assimilate his 'descent to the underworld' and so the unresolved shame he was left with became then the driving force behind the need to 'distinguish' himself from the 'lesser' men he was forced to mingle with, using the socially validated part of his identity to do so: his academic knowledge. The fact that he would characterize workers basically as dumb versus his educated self would transfer the shame from him to them, and in this sense, the whole system would be an attempt to pass on shame to other, less privileged souls who happen to have a lower social standing, substituting a healthy love of oneself, for power over others.
>
> Shame would be similar to the retrovirus and to Managerialism in that it subverts the order of means and ends: enhancing one's identity (the structure used to interact socially) becomes more important than living a fulfilled life (the end the identity is supposed to help fulfil). I wonder if this or something similar is present in people who adopt the doctrine of Managerialism out of the need to validate themselves, because shame might be lurking in their psyche.

The therapeutic tradition she draws on, Narrative Therapy (White 1997), de-constructs what it calls 'dominant narratives', unifying versions of what happened and why. Then it opens up a space to discover alternative stories, which also really happened. Their differences act like butterfly effects; new starting points which unfold over time into radically new possibilities in the future.

We apply Laura's therapeutic perspective not backwards, to long-dead Taylor, but forwards to new patterns to look for in the later history of management theory. Parts of Taylorism that were lost in the dominant history may still prove to be alive.

Larrikin antibodies

We can see something like this in the later history of Taylorism. Or, to use a different metaphor, the Taylorist retrovirus generated its own antibody. A continuous minority voice in management studies offered an alternative story. One was Elton Mayo, an Australian born in 1880, the year Ned Kelly died, just as Taylor's theories were forming. Mayo left Australia for the USA, where he became a Taylorist researcher at Harvard University. He began with a Taylorist study, which became a classic in organisation studies because it contradicted Taylorism precisely by trying to prove it.

Working with the Hawthorne factory of Western Electric in California, Mayo's team introduced a Taylorist reform of work practices, and found the predicted increases in productivity. Encouraged but not surprised they tried another reform, and again productivity increased. They continued to introduce improvements, and each time production increased. Finally they reverted to the original bad conditions. Contrary to the expectations of Taylorism, production improved yet again.

Mayo concluded that improvements must have been due at least partly to social factors, how members of the task group related to each other and to the researchers. In terms of scientific method, it was like the discovery of dark matter: careful linear analysis showed there was a decisive new factor not being taken into account.

From this remarkable paradoxical conclusion was born a new trend in Organisation Studies, the Human Relations movement. This emphasised social factors in the workplace, mostly ignored in classic Taylorism. Amongst these social factors, informal practices played a key role. Analysis solely of explicit, formal systems could miss key drivers of the system. If we want to talk of revolutionary new paradigms, this would be a good candidate.

Mayo has two larrikin features. We have associated informal systems especially with the Larrikin Principle. One precondition for Mayo's success was that he was able to challenge the prevailing orthodoxy, while staying basically within its terms. In effect, Mayo's work showed how the dominant model could be made to work in practice.

This introduces an ambiguity when judging him, the ambiguity intrinsic to the Larrikin Principle. Larrikins critique the dominant system, yet also make it run better. From one point of view, Mayo introduced a mutation into the Taylorist retrovirus, not a different line of theory. In this view, the line that followed his work continued basic Taylorist ideals, replicating them. It produced more productive *and* happier workplaces, just as Taylor claimed it would do.

Amongst those influenced by Mayo's project was Douglas McGregor (1960) who made the distinction between what he called Theory X, the dominant (Taylorist) tradition of linear control, and Theory Y, which looked for more complex, harmonious relationships in organisations. The term theory Y has largely disappeared from academic work in management, but all our management textbooks make some reference to it. It is managerial orthodoxy, part of what all management students learn as a basic truth of how to manage. Garratt's theory of Learning Organisations is a version of Theory Y that is now part of Managerialism, along with other similar schools. None of these call themselves Theory Y, but all incorporate its principles.

But what does this mean? Theory Y emerged in the 1960s, as a name for a potential revolution in management, building on Mayo's paradigm-breaking work. The revolution happened 50 years ago. If it were acted on, then or now, it would be a transformation of management and organisational life. Nonetheless, it has been absorbed into the dominant practice, so the revolution has not yet happened after all.

Worse, in a sense, it was already present, in a muted form, in Taylor's classic form of Theory X. Is it a New Paradigm? Or is it a tendency that is always present, always controlled, in Managerialism throughout its long history? Or is it both: needing a constant supply of larrikin energy to turn it into genuine practice, to convert the managerialist ideological complex into a blueprint for the change that it exists to resist?

The *Star Wars* guide to bad management

We believe it is no coincidence that three of the most popular films ever in the short life of cinema as an art form deal centrally with issues of organisation and management. In *Titanic* the ship is a triumph of technology colliding with forces that fall outside its arrogant linear vision. In *Jurassic Park* the triumph of technology in the service of the profit motive unravels, and literally produces its own monsters. *Star Wars* has the same mythological structure, addressing the same problem of society today: the dangerous contradictions that make up the managerial complex, the difficult issues facing everyone who works in any organisation.

Their villains are potent amalgamations of greed, power and linearity, symbolising aspects of capitalism and Managerialism. *Star Wars'* villain, Darth Vader, is one of the great creations of world literature. He is evil, of course, as villains should be. He is evil at three fractal levels. He is chief advisor to the Emperor at the galactic level. He is CEO to the Emperor as Chair of the Board of Death Star Inc. He is the absent, unknown father to the hero twins, Luke and Leia.

Vader is a psychopath by any definition of the term. Less obviously, so is the organisation he runs. The film allows us to see this fractally. The pathology recurs at three levels. In an analogous analysis, Canadian sociologist Joel Bakan provocatively assessed the psychological health of Corporations as if they were people. (2004) He applied the World Health Organisation manual on mental disorders to them, and diagnosed them as serious psychopaths.

Vader has all the qualities Bakan lists for psychopaths. He is also a cyborg, part-machine. Prosthetics give him power. He sees across the universe through a network of devices. He sends spaceships to do his bidding, armed with dehumanised Taylorist soldiers. Yet he is also vulnerable, only able to breathe with the help of a machine.

The Empire has almost achieved complete world domination, yet this is not enough for Vader. He built the 'Death Star' as an ultimate weapon, the final prosthetic extension of his self. It is a triumph of linear command-and-control thinking. Before, masses of dehumanised robot servants implemented commands from the centre. This single device will make even them redundant. It is the vision of a Taylorist manager with an unlimited budget.

Yet he is defeated, against all odds, by a small group of rebels, including his son Luke. The rebels use a plan of the Death Star to identify its one point of weakness. With linear planning, the object is no more and no less than the plan, so the rebels know their analysis of the plan will correspond to its reality. All complex structures contain contradictions, sites of weakness which cancel their strength. These two facts about linearity and complexity enable them to destroy the Death Star.

Or perhaps this is a butterfly effect. One aspect of the system was so minor that no one could have foreseen that attacking it would have had such catastrophic effects, in a system that was not seen to be at the edge of chaos. But it was.

There still needs to be a hero. Luke is a set of key contradictions: of noble birth yet raised in poverty, he needs to train his feelings, accepting the wisdom of the Jedi tradition, then learning to trust his trained feelings in the highly charged and complex moment of decision. Only this combination of discipline and freedom, intuition and reason, rebellion and dedication to justice can succeed.

The series fuses patterns of ancient myth with the modern problematic of organisations. The analysis captures the pathology of Managerialism, so close to Zadeh's Principle that we could name this the Death Star Corollary: all linear plans produce their own fatal contradictions. The solution they dramatise is the Larrikin Principle.

3. Coping with chaos and change

The other aspect of the New Paradigm is its response to uncertainty, change and chaos. As Clarke and Clegg show, the New Paradigm claims to have confronted and conquered the challenge of chaos. The New Paradigm within Managerialism seemingly incorporates chaos and complexity ideas so thoroughly that it is at home in the new environment. It does not need larrikins, or anyone else, to tell it about chaos.

In this section, we will look a little deeper, to see how Managerialism treats change, how far it has really changed itself.

Discourses of change

Change is recognised in management studies, as our management textbooks show. They all have a chapter on it, either alone or tied to 'Innovation'. But change in the textbooks is tied closely to what managers do about it. They must be 'change managers' or 'change agents'. This reduces change to a transitive verb, with managers as the subject, and organisations as the objects, of their managerial activity.

The perspective from below is different. Another larrikin, Bruce Watson, an education consultant, wrote a wry opinion piece for an Australian business magazine:

> You know the drill: when the going gets tough and problems abound in an organization, the top-level executive management team concludes that the staff must need professional development. Either that or another restructuring program is devised on paper, with very little thought given to the way the organization works (2004:65).

This piece, written for a popular audience, neatly captures a number of points about how discourses of change are seen from below. Watson is an informal ethnographer, identifying the core categories used by top-level management, as a distinct group with its own culture. For Watson, these senior managers understand their organisation as divided in two. The line is the same as Taylor's, though it occurs in a different place, between top managers and the rest, instead of between all (scientific) managers and the workforce.

In this model as Watson explains it, only the executive is active. They introduce change (variously called 'change management', 're-engineering', 're-designing', 're-structuring'). Staff are passive objects of change who are 'developed' or 'restructured'. The causality assumed is linear and uni-directional. If anything happens, it is because the executive makes it so.

All these are Theory Y measures. However, staff see them as mainly discursive: 'bullshit', to use a good larrikin word. The changes are precipitated by conditions outside ('the going gets tough') or inside ('problems abound in (the) organization'). Top management does not understand either. Taylorism provides them with no way of understanding the turbulent chaos of the external world. They probably created many of the internal problems.

Either way, management has no real solutions, but cannot show that. Mostly, Watson implies, the executive's solutions are not designed to grapple with the real problems. They have discursive effects, designed to boost the image of senior managers and reinforce their control of what counts as the truth. 'Discourse' in this sense has been influentially theorized by Foucault:

> In every society the production of discourse is at once controlled, selected, organized and redistributed according to a certain number of procedures, whose role is to avert its powers and its dangers, to cope with chance events, to evade its ponderous, awesome materiality. (1978:216)

As Watson implied, the management discourse of change tries to 'avert its powers and dangers' rather than understand it.

Foucault did not study change in organizations, but he did find similar patterns in discourses of sex and repression. He critiqued the widespread myth that sexuality was severely repressed in the Victorian Age, but now we are liberated because we can talk about it. Foucault argued for the need to go behind this myth, and the discourses which sustain it, to explore 'the will that sustains them, and the strategic intention that supports them' (1978:6). He noted that discourse does not simply reflect or control social practices. On the contrary, talk about sexuality flourished, while repression continued, managed now through an excess of talk.

Foucault did not explicitly use cybernetic terms or theory, but he essentially describes a cybernetic system in which talk (about sex in his case, change/chaos in ours) acts as a feedback loop designed to limit and control the object of that talk. Quiggin talked of actual restructures, not talk about them, as 'incessant'. In cybernetic terms talk and a kind of policy may go together, and serve similar functions.

Foucault's theory seems to imply that discourse as a feedback loop always dampens down change. In our cybernetic model even talk as a feedback loop may accelerate it. Conversely some material changes, e.g. restructures, as feed-back loops can dampen down the extent of real change. Yet, both discursive and tokenistic changes can also catalyse escalating series of changes. They can take systems far from the previous state of equilibrium, irrespective of what the change agent intended.

We can make a number of observations, based on this non-linear cybernetic model of organizational processes.

1. The conventional opposition between words and actions, which supposes that only actions have effects, is not true of complex cybernetic systems in far-from-equilibrium conditions. The power of words (ideas, representations) can be multiplied many times over in a cybernetic system, when allied with sets of actions that jointly form a system. This power can accelerate or resist change.
2. These tendencies of words and actions are not inherent properties of what is said or done, but functions of their place in complex cybernetic systems, themselves subject to change from time to time. Similar talk about sex or change that repressed it at one time might, on another occasion, even for the same person, be a stimulus.
3. Contradiction is not always fatal to a system and its long-term survival. It can be essential. For instance, pure, self-consistent Taylorism would have been untenable. Indeed, early forms had serious problems. But Taylorism contained, from the beginning, a series of contradictions to its dominant principles of linear control. Mayo's theory was a contradiction inherent in the original form, the contradiction that Taylorism needed to survive. Yet Mayo's counter-loop not only stabilized Taylorism. It will create instability if its tendency to humanize workplaces evades control.
4. Major changes in systems and contexts can arise from minor or unintentional initial changes propagating throughout a system. Zoë's analogy with the retrovirus system is a reminder that elemental 'lower' forms

can multiply at pathological rates, in systems without adequate defences. Taylor's basic ideas, without inbuilt constraints and correctives, spread wildly because they were simple and confused. If he had been a deeper, more scientific thinker, he would have had far less impact.

Resistance

'Resistance' is the other side of change. It plays a strange role in the linear ideology of Managerialism. On the one hand, it represents the reality that counters the efforts of 'change managers' to achieve change. On the other hand, this 'reality' is usually set in a linear fantasy world in which managers are always right, resisted only by the ignorant. Davidson's discussion is typical of how resistance is seen in management textbooks. 'Managers need to know why people resist change and what can be done about their resistance' (2003:419).

The taken-for-granted assumption is that managers' ideas for change are always good, and opposition never has any good reasons:

> Of course, a manager should not give up in the face of resistance to change. Although there are no guaranteed cures, there are several techniques that at least have the potential to overcome resistance (2003:421).

Just as Zoë saw Managerialism as a virus, so Davidson regards resistance as a disease, for which Managerialism has cures. That cure does not include listening and maybe learning. It does not include the possibility of reflection, to understand if the resistance is in opposition to change, or just to bad change.

Intelligent resistance is built into the larrikin response. Here is a typology of forms of resistance, illustrated from our larrikin archive:

i) Supporting the institution's goals against linear practices.
Jenny works in a section of her university dealing with student complaints. Her job is meant to make the students' lives better, and she does it with competence and dedication. Yet she feels that she and her unit are always battling against the linear dominant system.

> We're not popular, because it's not seen as core business, so we forever – forever! – have had to fight for survival, in what we feel is a very crucial role, yeah. And so we've always been anticipating. We've done it. That's why we're still there. You know, now – I don't know what happens in other sections, but in our section, we are well aware, that the enemy is out to get us. And so we are forever second guessing the enemy's strategies, so that we can counter them. That's what we do, we spend a lot of time doing that!

Jenny and her team are not valued by management for doing exactly what they are paid to do. As a larrikina she sees the conflict with senior management in a clear-eyed way, but she only opposes them in order to do her job properly. This consumes time and energy which could have been better deployed, but she feels she has to do it. They have better feedback and the more caring university that they need, but this happens in spite of them.

ii) Waiting for change to cancel itself out.
Tracey worked in the records department of her university, which introduced a new system three years previously, which never worked. Tracey and others in the office continued to use the old system:

> And it's a system that has just not worked, and we have ignored it, mmh? We have ignored it completely. And now, this system, they are going to change it for another one. Because it just did not work. I'm glad we ignored it! [laughter] It's awful, awful! And it is a service that the University bought for the whole University! To, to control all kinds of equipment, all sorts, even a table, even a tree!

Management clearly made a mistake in choosing this system. Mistakes do happen. But in this case it took three years to discover or recognize the mistake – not a rapid reaction-time. During this time, they were protected from the consequences of their poor decision by the larrikin resistance of Tracey and her mates. If management had used some of Davidson's tricks to overcome her resistance, they might have discovered their errors earlier the hard way, through the collapse of the system.

Tony, our larrikin bank employee, developed this into a strategy. Of his senior managers he says:

> They'll come up with an idea, two weeks down the track they will either change it or they'll forget about it, or whatever. So what you just say is, "We'll do it like this for now to the point where they're not asking us to do it like this anymore, and then we'll go back to doing it properly", sort of thing.

Tony is clearly the kind of employee that 'change managers' have in their sights, just as they are in Tony's sights too. This is a symptom of larger problems. Even so, his resistance and cynicism do not come from laziness or lack of commitment. He wants to do things the 'proper' way.

iii) Saving management from itself.
Tracey and Tony's resistance is small, local and without a major impact. We have stories in our archive where acts of resistance were played out for higher stakes, by competent managers who wanted to save the organization from its top bosses. Peter was a middle manager in a large NGO. He and his

fellow middle managers felt it was being run badly by a CEO/owner who was making irrational decisions. They got together as a group to plan what to do:

> But in the end we adopted the strategy – and when I mean we, I mean the middle management. We said "Well, basically, let's just do what we're told, so to push it to the top ones, and do something that's obviously stupid. And then, let's just let the shit hit the fan. We'll just implement it without question. We'll be like robots …
>
> And so this was the strategy, it was a Larrikin-type strategy. So we adopted this approach, where the persons at the top would say "Do this, this and this," and normally, we would have said "Well, you wanna think about that? I mean, that's gonna cost this, or, that's going to, the people in the field are going to react very poorly to that, because they've invested all this money here, or they've done this". Then we took this approach: "Ah? That's what you want? Without question we're gonna do it."

Peter and his fellows, competent and dedicated middle-managers, usually took on the larrikin role of providing frank, fearless advice to their boss. It was only after a long period of frustration that they felt they had to do something else. Paradoxically, their larrikin resistance took the form of no longer being larrikins, exposing the degree to which this organisation normally relied on their unacknowledged contributions.

The outcome, Peter reported, was positive. Opposition was mobilised to the CEO, who finally resigned. The company introduced a new management style which turned the situation around.

Total loyalty to a CEO is not always the best for a company. If there had been more larrikins in ENRON, for instance, it might have checked the fraudulent, unsustainable practices that led to the then greatest bankruptcy in US history (Hodge and Coronado 2006). In Australia, insurance company HIH collapsed after a long period when no one in the organisation felt able to tell authoritarian CEO Ray Williams what he needed to know about the flaws of his company (see Duarte, Gray and McAllister 2006). Managers and organisations everywhere need to learn the value of listening to resistance, not just trying to 'manage' it.

'Only the paranoid survive'

We have a complex, ambivalent attitude to contradiction. We see the multiple contradictions of the managerial Ideological Complex as harmful, yet larrikins and heroes of myth draw their power from contradictions. In terms of chaos theories, contradiction is not automatically a bad thing, to be removed wherever it is found. On the contrary, contradictions are everywhere. They are the products of dynamism and change, not to be ignored. They may be the best guides to the drivers of the system. In linear analysis contradiction is always fatal. In non-linear analysis it is sometimes harmful, sometimes not.

In this section we will sketch a more complex theory of contradictions which can interpret their meaning and deal with them, whether good or bad, benign or harmful. The key again is cybernetics. Here we use the influential work of cyberneticist Gregory Bateson (1971) and his work on schizophrenia.

Bateson argued that everyone is exposed to multiple contradictions. Normal people know this and have ways of managing them, so habitual they hardly notice what they are doing. This system is then held together by a loop he calls Meta-communication. This loop explains to participants how the system works, how meaning is to be made and interpreted.

People tend towards schizophrenia under conditions where they are exposed to continual contradictory double messages, delivered by someone who is loved and respected, with meta-communication being blocked. Victims of double messages are afraid or forbidden to acknowledge that contradictions exist, or to do anything about them.

There are two classic responses to this situation. *Paranoia* is the best known. Paranoiacs cannot sort good messages from bad, so they believe the worst, often as a hidden message. Another response Bateson calls *hebephrenia*, deliberate simple-mindedness. Hebephrenics refuse to believe in subtexts or hidden meanings. They are ideal workers for Taylorist managers. Unfortunately they are products of pathology.

We would be misreading Taylorism if we saw a purely linear theory, and denied or ignored its contradictions. It is these contradictions which have enabled it to survive as long as it has. There are good cybernetic reasons for this fact. The theory needs both positive and negative feedback loops to cope with the shifting dynamics of the environment. The longer it survives, the more contradictions it will accrete.

Paranoia has its defenders. Andy Grove, highly successful CEO of Intel, used the slogan 'Only the paranoid survive' as the title of a popular book (1996). Andy Grove's slogan contained its own contradiction, plus an awareness of that contradiction (that he, the CEO of a hugely profitable company, was calling himself paranoid). His book explicitly connected a potentially paranoid consciousness with a successful New Paradigm response to crisis and uncertainty. Under his 11-year reign as CEO, Intel's worth went from USD 18 billion to USD 197 billion. Not all self-confessed paranoiacs make so much money, but it is a good advertisement.

Andy Grove does not really doubt his sanity. His slogan provokes an analysis of how contradiction functions in extreme situations, in genius, madness or larrikinism. Bateson makes some useful distinctions. Crucial is the meta-communication loop that is indispensable for a system (individual, organisation) to cope with contradiction.

This function can be carried out by a range of critics, one or many, 'speaking truth to power' to and for the non-powerful. The Larrikin Principle is a way to do this. One key larrikin function in an organisation is to demystify communications, especially from above.

Bateson's model helps make sense of the contradictions of Managerialism. Communication becomes dysfunctional when contradictions are unlabelled, especially when they come from respected authorities. We repeat: contradictions as such are not the problem. It is the refusal to recognise or even think about them. Those who are unaware of contradictions or block them from consciousness will be paralysed and incapacitated. In the interests of their mental health and their capacity to function at the highest levels, they need critical pedagogy, and a larrikin within.

Managerialism

Managerialism is one name we gave the antithesis of the Larrikin Principle, yet in some ways it is the hardest of enemies, because it is so hard to pin down. It is a fine example of an ideological complex, a constantly shifting set of contradictions accommodating all criticisms in order to neutralise them. It is like Proteus in Greek myth. Odysseus the trickster tried to make this sea god tell him the truth. Proteus kept changing his shape. But Odysseus knew that if he kept holding on, Proteus would finally reach his true shape, and reveal all.

In this chapter, we have wrestled with Protean Managerialism until he ran through his repertoire and was left speaking his surprising truth: linear management is no way to run a business or a world, though Managerialism will never admit that. Alternatives are well enough known already to begin the revolution that Managerialism exists to block. The Larrikin Principle encompasses two contradictory principles, both as necessary as Odysseus' two arms with which to grasp Proteus. One is paranoiac suspicion of every managerialist claim to do good. The other is naïve belief in the literal truth of these claims, and the determination to make them happen.

Chapter 4

Neo-Liberalism and its disconnects

This chapter should be easy to write, in the wake of the global financial crisis of 2008. If our only task was to criticise Neo-Liberalism, this has surely been done already, by leaders across the globe as eminent as US President Obama, UK Prime Minister Brown, French President Sarkozy and Australia's Prime Minister Rudd. The task is so easy that, ironically, it becomes hard. How can we justify saying again, as obscure academics, what many others have said more eloquently and with greater authority?

Yet we have some warnings, before the dance of triumph ends. If we are right, Neo-Liberalism is not a single, coherent doctrine that has now been demonstrated to be wrong. On the contrary, it is an ideological complex, a functional set of shifting contradictions, designed to adjust to new conditions, to offer new justifications for the interests it exists to defend. If one part of a contradiction is attacked, the other part comes into play. A mere collapse of the world financial system will barely dent its credibility, as other parts of the complex kick in.

'Neo-Liberalism' as an ideological complex is like one of the heads of the Hydra, the many-headed monster of ancient Greek mythology. Hercules, strongest of men, tried to cut off its head, but every time he did so, two new ones grew in its place. The Hydra was a non-linear monster, responding paradoxically to a sensible linear hero's attempts to kill it. To defeat it, Hercules had to cauterize the wounds with a firebrand ('Hydra' comes from the word for water, so Hercules used its opposite). Finally, he cut off the one head that was immortal.

There are many contradictions that we need to identify when seeking to neutralise the hydra of Neo-Liberalism, as an ideological complex and as a part of a larger complex that includes Managerialism, itself an ideological complex that is hard to skewer, as we have found.

We will summarise some contradictions here, before going into detail in the chapter that follows. 'Neo-Liberalism' in its heartland is a name used mainly by its enemies. None of our management textbooks mentions the word. President Bush did not call himself a Neo-Liberal, in spite of being seen by others as its champion. He preferred different terms, like 'free trade', 'democracy' and 'pro-business'.

Prime Minister Kevin Rudd did not use the word much himself either, until his recent conversion, when he named it as the system that has failed.

'Neo-Liberalism' as a term builds in a repudiation in advance, making it harder to attack. 'Neo-Liberals' can easily say: 'No, I'm not that'.

The key contradiction of 'Neo-Liberalism' pivots around the word 'liberal'. 'Liberal' means generous, kind, nice. But 'Neo-Liberal' does not mean a new (neo-) way of being generous or nice. On the contrary, it goes back to the profit motive in traditional capitalism, to old ways of being nasty and mean in the pursuit of profit. 'Neo-' here means something like 'not-' (liberal). Yet 'Neo-Liberalism' refers to policies and practices that still claim to be 'liberal', in the older sense: both Liberal and not-Liberal.

If we were Hercules our heads would already be spinning, but there are more contradictions. Neo-Liberalism and Managerialism describe the same system, the currently dominant form of capitalism, at different fractal levels, yet in some respects with opposite effects. 'Neo-Liberalism' emphasises freedom (for business) from controls and regulations. Control is a core aim of Managerialism. Yet the contradiction between levels is repeated within each level. Neo-Liberalism as a policy is applied prescriptively to governments, in the name of freedom of the markets. In this respect it is a managerialist device used by a conglomerate of big organisations to manage the globe. Conversely Managerialism, as we have seen, has its softer side.

There is a final less obvious contradiction, at the highest fractal level. 'Globalisation' is a word commonly used in sociology and business to describe the current state of the world system, the ultimate context of every business and social activity. In business, it is often used as if it were almost the same as Neo-Liberalism. In this sense, globalisation is the product of Neo-Liberal activities, a condition so congenial to Neo-Liberalism that no other way of acting could hope to work.

Yet globalisation is a challenge faced by every firm. Globalisation is not an easy extension of Neo-Liberalism and Managerialism in order to become the basis of best practice, but the contrary. Globalisation looks more like master than servant, vast, intractable and chaotic. Whatever it is, we need to distinguish it from the Neo-Liberal, Managerial Ideological complex.

1. The many faces of Neo-Liberalism

Our way of opening up this theme is to continue to follow Bob's journal. This time he reports reactions to the Mexican press on his arrival.

> Two main stories dominate the media. The main one is the upcoming budget exercise, which Neo-Liberal President Calderón calls an 'integrated fiscal reform'. The other is the issue of electoral fraud in the presidential elections of 2006. This theme must be getting legs now, in mid-2007, because the anniversary of the election is coming up. I'm interested to see how Calderón's budget plays out, as a display of the Neo-Liberal credentials he campaigned

on, and how the disputed election is a factor, one year on. I'm not a Neo-Liberal, but I come from a country which keeps electing a government which proudly proclaims its Neo-Liberal affiliations. Does that make me complicit with Calderón and his policies? I don't think so! (From Bob's journal, 17 June, 2007)

Our interest here is not just in the content, the Neo-Liberal policies of this Mexican government, but how this is framed in Bob's consciousness, as a typical citizen from the 'developed' world.

Coming from a 'first world' country like Australia he has the shock of seeing the word 'Neo-liberal' used up-front, as was not done in his own country though at that time it followed these policies without acknowledging the word. Six months later the Rudd Centre-left government would be elected, with Rudd calling himself an 'economic conservative'. Next year Rudd launched his attack on Neo-Liberalism itself, and the name was speakable in Australian politics.

Yet when Bob wrote this entry, he did not know the future that awaited. He inhabited two worlds which he could not relate. In Mexico he was immersed in Mexican discourse about the world seen from a Mexican perspective, mediated through Mexican media. Although as an electronic citizen he could and did read the same media from Australia, there was something schizogenic in being surrounded by it. It made the view from Australia, at a distance, filtered through English-language sources, seem quaint and unreal.

That is an experience we recommend to all citizens of English-speaking 'developed' countries, managers as well as others. These filters have effectively prevented most from understanding the realities conducted under this term that they hardly know or use: Neo-Liberalism.

To look at it from the other end of the telescope, it is salutary to look at another term not used in management textbooks or discourses: 'the Washington Consensus'. This term originated in one article by one policy analyst who did not expect the fame and controversy that ensued. After decades of policy work, mainly associated with Latin America, John Williamson decided to put down in writing the implicit rules that seemed to govern conditions attached to international loans provided to 'developing' nations by key Washington-based institutions which in practice regulate the world economy: the International Monetary Fund, the World Bank, and the US Treasury being among the most important (1990).

Williamson abstracted 10 factors shared by the main funding agencies and US Federal authorities, which he said represented a consensus amongst Washington-based authorities. The first condition is fiscal discipline. 'Washington believes in fiscal discipline,' he says, though he wryly notes that it does not practise it. The deficit should be achieved by cutting expenditure rather than increasing revenues, especially subsidies.

Yet one avenue of expenditure is praised: investment in health and education, 'in a way that will benefit the disadvantaged'. Other policy directives, such as privatisation, tax reform (to 'broaden the base') and 'trade liberalisation' form key parts of the package. Williamson insists that these measures are not ideological, a moral injunction to purity. Rather, the advice is intended to further the self-interest of the countries to which it is directed (although not necessarily with a weighting of the interests of the constituent classes identical to that of the ruling elite in those countries (1990:17).

Williamson's short essay may have had its impact due to his having a touch of the larrikin in his style. He was a Washington insider spilling the secrets of the club, reporting the obvious but unsayable, inserting criticisms in a dry, understated wit.

In spite of his later commentaries, we are not sure what Williamson intended by this article, how far he was a hidden larrikin, an Insider like Jeffrey Wigand, aiming to provoke precisely the outrage that happened. In the article itself he said:

> The paper is designed to elicit comment on both the extent to which the views identified do indeed command a consensus, and whether they deserve to command it (1990:2)

'Elicit' is a weak word for the explosion of comment that followed, especially from critics of Neo-Liberalism. Whether or not he meant to provide a clarifying target, that was exactly what he did. A long line of critics, like distinguished US linguist and activist Noam Chomsky (1999), seized on the term and doctrine, as Williamson expounded it and climbed in, without pausing to thank Williamson.

Williamson delivered his paper in 1989, looking backward not forward. Few now defend the Consensus policies as benefiting nations like Mexico, or Brazil's more complex case (Williamson 2008). In a later response he acknowledged that the term is now 'a damaged brand' (2003:1). Joseph Stiglitz, Nobel Prize-Winner in Economics in 2001, former Chief Economist of the World Bank, delivered a crushing assessment of the effects of Washington Consensus policies (2002). The World Bank and the International Monetary Fund have officially upgraded poverty reduction as an aim. But as Williamson noted, they were already claiming to do that in 1989.

Living with Neo-Liberalism in the developing world

If the Washington Consensus, like Neo-Liberalism, is officially discredited and dead, it is salutary to notice what the other heads of this Hydra are still doing. Williamson's list is a useful yardstick for looking at Calderon's budget. We begin with the media presentation, since that is what had immediate effects, both on Bob as visiting larrikin observer and on the Mexican public.

President Calderón and his Finance Minister claimed to follow some of the principles favoured by Neo-Liberalism. One aim, they said, was to achieve a balanced budget. Equally Neo-Liberal in spirit was a proposed tax reform, promising to attack the informal sector to ensure greater compliance. They added a reform so innovative that even architects of the Washington Consensus had not thought of it: a new 50% tax on aerosol paint sprays, to hit graffitists in the informal media sector. This would not raise enough revenue to register on a Washington-Consensus meter, but it would feel good to the suits.

But in his first media blitz, President Calderón insisted that the heart of the reforms was not Neo-Liberal at all, but a passionate attempt to shift the balance between rich and poor. On 18 June, speaking 'from one of poorest municipalities in the country', according to Left-Wing periodical *La Jornada*, Calderón vowed that, at last, public resources would pay for 'the enormous social debt' with Mexico's poor (19-6-07:12). Two days later he said: 'This is the last chance for Mexico to make a change in the war on poverty'. He christened his initiative 'a budget reform for those who have the least'.

He assured his audience that, if this budget were to be approved, 'the population in extreme poverty would drop by 30%, and Mexico would be a leader in Latin America for constructing infrastructure' (21-6-07:10). This would be done via a new tax on businesses, which would fund dramatic initiatives to help the poor. Neo-Liberal President Calderón playing Robin Hood, robbing the rich to give to the poor?

As days passed, new analyses emerged to cast a different light on the measures. In an editorial, *La Jornada* (21-6-07:2) noted that the amount to be generated by the reforms was almost exactly the amount needed to maintain Mexico's debt repayments up to 2010. Who could doubt that a fiscally-balanced, outwardly orientated Neo-Liberal government would give this top priority, as Calderón's predecessors had done? If so, little would be left over for the poor in Calderón's great struggle on their behalf. Nothing in the budget guaranteed anything to ease these doubts.

From the Right, other queries were raised. Enrique Quintana in *Reforma* (20-6-07: Negocios 9) saw the measure driving up unemployment. Invex financial group showed that, of the top 20 big companies, only 4 would have to pay more than they do now (Gonzalez *La Jornada* 24-6-07:3). The 'tax on the rich' would hit only small to medium-sized enterprises (SMEs). Carlos Fernandez-Vega (J 21-6-07:28) doubted that the informal economy would be much affected. He saw it as likely to become just another reason for a *suborno*, a bribe, among many that already insulate the informal system from contributing very much to general revenue.

A casual visitor to Mexico from the first world might say to all of this:

Ho-hum. So Calderón is a hypocrite? He's a politician! What do you expect? It is true that the pattern is recognisable, but that only makes it more important, precisely because it is so transnational. Neo-Liberalism is not a pure package of policies, which this individual politician failed to implement fully. As Williamson noted back in 1989, this 'failure' is built into the package. The ideological complex has always included positive rhetoric along these lines. The details change according to the situation, but are never meant to be acted on.

Neo-Liberalism down-under

Meanwhile back in Bob's Australia, Neo-Liberalism reigned supreme. Or did it? As he was writing, a swing was already occurring away from the incumbent Centre-Right Neo-Liberal party led by John Howard towards the Centre-Left Labour Party led by Kevin Rudd. Rudd won the 2007 election and denounced Neo-Liberalism in 2008, as we have seen. But in 2007 he still called himself an 'economic conservative', code for 'Neo-Liberal'. Australia's status as a Neo-Liberal country is complex and uncertain.

On the one hand, as a rough generalisation, we can say that Neo-Liberal policies were supported for 25 years by parties supposedly of both left and right, by the Labour Party, in power between 1983 and 1995, and the conservative Liberal-Coalition, between 1996 and 2007. Yet when we scratch below the surface the picture becomes more confusing and messy, in ways that say much about Neo-Liberalism itself.

As we said, 'Neo-Liberalism' as a term hardly existed in Australia. Its opponents mostly called it 'Economic Rationalism', a term popularised by a sociologist, Michael Pusey (1991). In a commentary on the short but twisted life of this concept, John Quiggin noted that it began as a term used positively by a Labour government, in the early 1970s, to describe their own policies (1993:277). Later the term was associated with 'rationalisation', a euphemism for cuts. This made it negative, for those on the receiving end. It also made a direct connection with the same policies as applied within firms or organisations, which 'Neo-Liberalism' does not. This has the merit, for a holistic and fractal analysis, of connecting policies Washington applies to a complex external world to decisions taken by governments for their citizens.

Pusey called it a 'locust strike', a plague which had taken hold of Australia's policymakers with dramatic speed (1991:1). In his book, based on extensive interviews with Canberra public servants, he found a Canberra Consensus, a cousin of Washington's. This mainly took hold under Labour governments, headed by two politicians often called 'larrikins', Bob Hawke and Paul Keating. Pusey has critics, but no one doubts his claim that these policy settings were indeed dominant, both in the public service and in the political system.

We have no intention of covering 25 years of Australian policy here. It is marked by contradictions which complicate all simple generalisations. We illustrate the complexity with a single policy issue. In 2005, the Conservative Prime Minister, John Howard, pushed through what was regarded as an extreme Industrial Relations (IR) reform which aimed to dismantle the rights and protection of workers. It was seen on all sides as Neo-Liberal policy, without the name being used.

However, even those who supported this Prime Minister or this policy were not pleased with what they got. Right-wing economics editor Alan Wood wrote an article headed: 'IR bill isn't ideal, but it's going to work'. (*Australian* 2-11-05). He did not explain why it was going to work. It was easier to say where it was not ideal. 'John Howard's Work Choices Bill, to be introduced in parliament this morning, will fail one simple but effective test of legislation: its length. It will be slightly more than 700 pages' (2005:12). He adds:

> There is no escaping the fact that this is an undesirably complex piece of legislation, nor the irony that it is being introduced by a Prime Minister who only a few weeks ago announced a taskforce to reduce the regulatory burden on business.

He gives his diagnosis: government distrust of the relevant statutory bodies.

> As a consequence of this distrust, federal legislation has been drafted to try to cover every contingency and regulate in the minutest detail every possible decision-making process.

Wood regrets this contradiction, but it is no coincidence. It goes to the heart of Neo-Liberal policy and managerialist ideology. On the one hand is the rhetoric of freedom, removing controls, while on the other the inexorable will to control produces disorder and a failure of control. Zadeh's principle applies here, to legislation by the state, exactly as it does to lower levels, to organisations and groups.

In complex conditions, attempts to be precise lead to irrelevance or meaninglessness; in this case, over 700 pages of legislation. Even Wood, supporter of the principle of the legislation, called it 'another bugger's muddle'. The legislation was passed, but the backlash was intense, fanned by the still Neo-Liberal opposition.

By September 2007, just before the election was due, Prime Minister Howard faced an 18-point deficit in the polls, attributed by most to the unpopular IR bill. As one of his disappointed supporters wrote, in a letter to the editor:

> The Prime Minister refuses to recognise that Work Choices is the problem. The Liberals have attacked the fundamental building block of Australian

> society – the fair go. The public will not accept this legislation in sufficient numbers to re-elect the Coalition… With no sign of the Government recognising the bind it's in and doing something about it, the champagne is ready for popping at Labour Party headquarters. I, for one, won't be raising a glass. Michael Stanbridge, Letters to the Editor, *Australian* 5-9-07:19).

Michael Stanbridge proved a good prophet, though by then prediction was easy. Howard suffered a landslide swing, losing his once-safe seat in the carnage. What is interesting is Stanbridge's reason for this loss. The Larrikin Principle of 'the fair go' turned out to be an election-turner in Australia. How could this be, after 25 years of Neo-Liberal 'Economic Rationalist' consensus?

Ray commented on Neo-Liberalism, even using the term, to explore its consequences for the Larrikin Principle in Australia today:

> I'm not sure that [larrikinism] is as strong an element in today's Australia as it might have been 20, 30, 50, 100 years ago. I think we're becoming a very, how should I say, very bureaucratized, very shaped, very conformist society compared to what we used to be like, which may have something to do with, it probably has a lot to do with the acceptance in the Neo-Liberal ideology, both in terms of how the economy operates and by implication, both in the way that people operate in their own minds – because that kind of thinking, that Neo-Liberal thinking seems to be coming into the way that people think about their actions.

Ray sees larrikinism as waning in a changing reality in which Neo-Liberal ideology is increasingly dominant. He makes the connection between different fractal levels, from the lowest ('in their own minds') to the level of the actions of the state.

Ray is an intelligent, well-informed larrikin (sceptical, outside systems of power) who criticises larrikinism. But like Michael, he sees both principles in play. Each co-exists with the opposite, to a degree that was not clear to either man. For Michael, a Coalition supporter, his party should have combined Neo-Liberalism with larrikinism. Evidently for Michael, Howard achieved that balance or contradiction in the rest of his long reign. It seems that in Australia contradiction rules, not Neo-Liberalism on its own. Co-opted larrikinism is a crucial part of this dominant contradiction.

Contradictions and the ideological complex

Neo-Liberalism is impossible to define crisply because it is an ideological complex. No single, coherent definition underlies its different uses, as linear practices of definition like to assume. There is only a continual wavering between terms of a contradiction, as the powerful seek to impose upon and persuade others. They express and exercise their power through linear models which they know will alienate those they seek to rule, while presenting that power in benign, positive terms as that which the powerless really want.

Even agents of the Washington Consensus, the policies actually used to administer Latin America from Washington, expressed concern for the poor. To achieve its fiscal aims the Consensus blurred the interests of the country and the interests of ruling elites. As we saw in the previous chapter Taylor had a comparable Ideological Complex. Key to his success was his claim that everyone, bosses and workers alike, would benefit from these best-practice policies, and be grateful.

As larrikin analysts we note that these 'beliefs' were not translated into practice. They appeared mainly in words, in rhetorical justifications. Since the justifications enabled them to pursue their linear policies with greater freedom, they can be seen to have a cybernetic function. They are a dampening feedback loop, which counteracts resistance to Neo-Liberal policies.

The term Neo-Liberalism was founded on a contradiction, as Chomsky noted:

> The term "neo-liberalism" suggests a system of principles that is both new and based on classical liberal ideas: Adam Smith is revered as the patron saint... The doctrines are not new, and the basic assumptions are far from those that have animated the liberal tradition since the Enlightenment (1999:19)

To illustrate Chomsky's critique, Neo-Liberals hate 'Liberals', calling them long-haired pinkos, 'bleeding hearts', as soft on crime as on victims of harsh economic policies. Neo-Liberals are Anti-Liberal. Yet, they base their right to use the word on claiming links with 18th century 'liberal' philosopher and economist Adam Smith.

This is not just a slippery use of terms, to be corrected by more accurate ones. The slipperiness allows a whole set of contradictions of the ideological complex to seem like consistent parts of a coherent position. The core contradiction concerns the relation between policy and economic theory. Milton Friedman, the US economist, won the Nobel Prize for economics in 1976 for his contributions to monetary theory, but this work hardly figured in his influential manifesto for Neo-Liberalism, *Freedom to choose* co-authored with his wife Rose (Friedman M and R 1980). Instead he used his prestige as an economist to claim that economic freedom, understood as maximum freedom from government regulation, was the only effective basis for political freedom. He backed these claims up with an underwhelming array of empirical evidence, plus the obligatory appeal to Adam Smith.

Smith was indeed a 'liberal' in his time, though he did not use the word. He favoured freedom against the arbitrary and inefficient exercising of power by monarchic governments and powerful commercial groups alike. He was as suspicious of the motives of powerful economic interests, the MNCs of his day, as of monarchs. He did not share Friedman's naïve claims that only governments fail, that markets never do.

He also foreshadowed chaos theory with his idea of an 'invisible hand' in economic life, where countless individual acts of self-interested choice could aggregate into a benign system. Talking of the decisions of individual consumers he said:

> He is in this, as in many other cases, led by an invisible hand to promote an end which was no part of his intention. Nor is it always the worse for the society that it was no part of it....
>
> I have never known much good done by those who affected to trade for the public good. It is an affectation, indeed, not very common among merchants, and very few words need be employed to dissuade them from it (IV. ii, 2007:312).

In terms of chaos theory the 'invisible hand' foreshadows the idea of emergent structures, bringing a better order out of chaos than monarchs and other planners could.

Stiglitz, branded a liberal by the Neo-Liberals, argued that Smith's brilliant insight is only true under certain conditions:

> It turns out that these conditions are highly restrictive. Indeed, more recent advances in economic theory – ironically occurring precisely during the period of the most relentless pursuit of the Washington Consensus policies – have shown that whenever information is imperfect, and markets incomplete, which is to say always, *and especially in developing countries*, then the invisible hand works most imperfectly' (2002:73, his italics).

He argued in favour of some regulation to correct the imperfections of the market. In the wake of the 2008 financial crash, this is now a consensus view, even in Washington. This does not repudiate Smith. It takes his discovery of non-linear (dissipative) complex systems from an article of faith or a dogma to a scientific hypothesis to be tested and refined.

Neo-Liberals operating the Washington Consensus have aggressively intervened in the markets and politics of other countries. They used linear analysis and proposed one-size-fits-all linear solutions, claiming good intentions, with the lack of success that Adam Smith would have predicted. As Williamson points out, they were unable to intervene on this scale in their own economy. American power enabled them the freedom to devastate the economies of the 'developing' world, including Latin America. That was not the kind of freedom Adam Smith had in mind.

Fractals of the Neo-Liberal complex

Neo-Liberalism is an Ideological Complex at the international level, the highest fractal level. Calderón is influenced by it, and also repeats its forms at a lower fractal level, Mexico. At neither level is it a coherent set of policies and positions. It is a functional combination of the same set of contradictions, which do not cancel each other out but serve the interests of one side, Big Capital. Yet, at both levels, the ideal pattern is distorted by pressures from the context, as will always be the case. Ideological Complexes exist precisely to battle these pressures, and are always distorted in this way.

In this case Calderón made sure he did not offend major industries, but his measures did affect smaller capitalists who formed part of his power-base. The contradiction remains not only between rhetoric and action, it also exists to some degree at the level of policy. Most major firms who benefited from Calderón's budget were foreign-owned multi-national corporations, such as Wal-Mart, but some were Mexican-owned. Calderón's base group consists of major national and international investors. His opponent in the election, Andres Lopez-Obrador, claimed:

> A century ago, some 300 families were the rich, and controlled the government, now there are less than 100 (*La Jornada* 23-6-07:6).

Whatever the numbers, it is not in doubt that there is massive continuity between pre-revolutionary and current elites, and thus between neo-colonial regimes from the past and the present 'Neo-Liberal' ruling group.

Ideological complexes repay further fractal analysis. Major multi-nationals in Mexico like Wal-Mart repeat the Neo-Liberal and national ideological complexes, claiming to benefit Mexico at the same time as ensuring maximum profits, doing so through a similar set of contradictions. Stiglitz uses Wal-Mart as an example of the dangers of global actions opened up by Neo-Liberalism, exacerbated by the normal practices of MNCs. He begins the story within the USA. When Wal-Mart comes into a community, he says, local shopkeepers fear they will be swamped by Wal-Mart's buying power. They are right. But after local competitors are driven out, Wal-Mart raises its prices to previous levels. 'These same concerns are a thousand times stronger in developing countries' writes Stiglitz (2002:68). He attempts a balanced perspective:

> Although such concerns are legitimate, one has to maintain a perspective: the reason that Wal Mart is successful is that it provides goods to customers at lower prices.

The case of Calderón's deceptive budget shows a catch with this position. If national governments give big multinationals like Wal-Mart large tax breaks, they do not have to be more efficient to provide goods at lower prices. A discriminatory government subsidy is all they need. The Washington Consensus opposes subsidies, but only to local firms. Massive incentives to foreign interests are OK. Taxes from local citizens support the destruction of local industry. At the same time they make MNCs look good without having to try.

All this is a normal outcome of Neo-Liberal policies carried out by Neo-Liberal national governments. It is yet another paradox of the ideological complex. A rhetoric of competition is managed by the secret if not invisible hand of an alliance between Big Business and ordinarily corrupt politicians. The result is less competition, less efficiency.

2. Globalisation

Management textbooks see globalisation as important, yet commonly give it a distorted, truncated sense and history. There is an uncertainty which reflects deeper contradictions of the Managerialism complex. The problem is that Managerialism claims an edge in its grasp of the global scene, the site and proof of its universal adequacy, yet the hyper-complexity of the global world is precisely what its linear models cannot cope with. It resolves the problem through rhetoric, and political levers of Neo-Liberalism to rig the global game.

The Davidson discussion of globalisation is typical of the textbooks: 'Globalisation was initially conceptualised as the worldwide process of economic and industrial restructuring', they begin. They then add: 'Today, globalisation is also seen to include the process of continuous change to gain competitive advantage' (2003:146).

This gives it only a short history. 'Initially' implies it began in recent times, after 'economic and industrial restructuring' happened. 'Globalisation' in this view is mainly about corporations and what they do. Complexity and growth come only from competition on a global scale, instead of within a nation state as before.

Our students are regularly amazed when we introduce them to Marx and Engels' words, written in 1847, 100 years before globalisation was 'invented' by Neo-Liberalism:

> The need of a constantly expanding market for its products chases the bourgeoisie over the whole surface of the globe. It must nestle everywhere, settle everywhere, establish connexions everywhere.
> The bourgeoisie has through its exploitation of the world-market given a cosmopolitan character to production and consumption in every country...
> In place of the old wants, satisfied by the productions of the country, we find new wants, requiring for their satisfaction the products of distant lands and climes (1970: 38–9).

Marx and Engels' analysis of the motives and processes of globalisation is impressively current. Even they were not the first to notice globalisation. In the 18th century Adam Smith entitled his great work the *Wealth of Nations* because he saw exchanges across national borders as the crucial new factor in economic explanation.

Marx does not call this process 'globalisation'. The process he describes was already 300 years old when he wrote. He only seems a remarkable prophet to those who believe, like management textbooks, that globalisation was invented in the 1980s. Nor does he use the term 'non-linear', but like Smith he clearly grasped it. These two thinkers have different but complementary perspectives on the complex system they co-discovered. Smith's 'invisible hand' produced good outcomes from the everyday actions of ordinary citizens. Marx describes a pattern of exponential growth generated by the logic of capitalism, driven by specific actions and practices, but increasingly out of control.

In an influential recent book Harvard academic Michael Hardt combined forces with Italian political theorist and activist Antonio Negri to take the Marxist story into the age of globalisation. At the time, Negri was an inmate of Rebibbia prison in Rome on disputed charges of 'terrorism', a critic of state power like Ned Kelly, as well as a victim, like him, of unjust laws and processes. Hardt and Negri summarise the crucial stage in the formation of the contemporary patterns of globalisation after the Second World War:

> The new global scene was defined and organised primarily around three mechanisms or apparatuses: (1) the process of decolonisation that gradually recomposed the world market along hierarchical lines branching out from the United States, (2) the gradual decentralisation of production; and (3) the construction of a framework of international relations that spread across the globe the disciplinary productive regime and disciplinary society in its successive evolutions (2000:244–5).

Like management writers on globalisation, they see something new happening after 1945, with roots in earlier periods. We agree with their analysis into three mechanisms, but we stress the tensions and contradictions implied in this story.

'Decolonisation' seemed to reverse the relentless progress of bourgeois imperialism as Marx described it. But Hardt and Negri see it as its opposite, another moment in the strategy of empire to impose its rule on those who fought to be free. Mexico and Brazil are two countries among many whose 'liberation' from their colonial masters, Spain and Portugal, left ordinary people still dominated by local elites.

'Decolonisation' did merge into its opposite, 'neo-colonialism'. Yet this is not the whole story. Decolonisation was not the idea of ruling elites, even if they tried to coopt it. 'Decolonisation' is more fuzzy and shifting than Hardt and Negri make it seem.

Decentralisation also involves contradiction. It can be a strategy for the new order because there are now more effective mechanisms to hold this new form of the world together. Yet decentralisation still threatens Managerialism's linear will to control. Decentralisation too is fuzzy and unstable.

Control is still the core theme of this new stage of Managerialism. This is clear in their third point. 'Disciplinary productive regime' is their term for what we call Managerialism. They date the globalisation of Managerialism well before 1945. They point to the paradoxical popularity of Taylorism with the socialist regimes of Lenin and Mao (2000:248). They could have mentioned the equally great popularity of Fordism with Hitler. 'Globalisation' was the medium through which linear Managerialism propagated and mutated, like the retrovirus that Zoë Sofoulis called it. It was not the retrovirus itself.

Gurus on globalisation

In business as in social theory there is agreement that globalisation today is new and challenging, though not what it is. At one extreme, voices talk of a change so radical that the term 'paradigm' (shift or change) is often used. We looked at Clarke and Clegg's account of this development in management (1998). In this section we see how the idea applies to phenomena associated with globalisation.

One writer from the world of business who influentially theorised this shift is Austrian-born US management guru Peter Drucker. His concept 'Management by Objectives' (MBO, 1954) is orthodoxy in all management textbooks. He also wrote on the globalisation revolution.

His *Post-Capitalist Society* (1993) begins with a vast, though brief, historical sweep. He classifies previous history into the familiar stages. He then claims that a new era has now begun: post-capitalist society. This sounds challengingly radical. How has capitalism been overcome, without the revolution Marx thought necessary? What was Drucker's revolution?

He describes the decisive change in Marxist terms, a new 'means of production', i.e. knowledge (1993:7). This elevates 'knowledge workers' to the pinnacle of this new form of society. These are not 'intellectuals', who will not find an easy place in his new world. They are managers.

One way to describe Drucker's grand radical vision is as a flood of MBAs with MBOs in their briefcases taking over the world. He dates this revolution from 1945, as do the management textbooks, but he is clear who began it. Frederick Winslow Taylor was the hero, vilified and misunderstood in his time, vindicated by history (1993:32).

Drucker's management theories shared Managerialism's contradictions. Social concern softened the impacts of basic linear strategies in the interests of greater control. As a theory of revolution and an account of globalisation it falls short of the rich complex story of Marxists and other social theorists on globalisation. His 'revolution' is just Taylor's managerialist fantasy, managers as the solution to all conflicts and problems of the world.

All management textbooks agree that a decisive new factor in globalisation has been the development of the Net. As US business academic Helen Deresky wrote in her influential international management textbook:

> Of all the developments propelling global business today, the one that is advancing the international manager's agenda more than any other is advances in **information technology (IT)**' (her emphasis 2002:8).

There is no doubt IT has involved a momentous change. The electronic sphere is highly globalised, and very fast. As Castells (2000) pointed out, information travels around the planet at the speed of light. Yet it is not so clear that business has a unique or adequate grasp of this new technology.

Certainly US entrepreneur Bill Gates became one of the richest men on the planet because he had a successful business model for exploiting this commodity. That does not mean that he had a unique insight into the technology itself. As a business guru he promoted a package that combined IT technology (already available from Microsoft) with elements from chaos theory in *Business @ the speed of thought* (2001) but his basic business models are still close to Managerialism.

Czerniawska and Potter wrote their vision for business's virtual revolution:

> The use of information is a revolutionizing force: enlightened companies have realized that the trick of winning this battle is to hold off converting their product into its physical form until the last possible moment. Information – unlike physical goods – can be transmitted round the world literally at the speed of light; it can be manipulated with computers far faster than any physical object can be changed; it can be duplicated at no cost, whereas reproducing physical goods is a slow and expensive business. It makes sense, therefore, for as many activities as possible – product development, sales, marketing and distribution – to be performed in the virtual world, thereby increasing speed and decreasing costs. (1998:7)

This sounds impressive. Their view of the cybersphere is inclusive. Microprocessors and other electronic forms are embedded in other material processes. But the apparent solution of all material problems rests on a rhetorical trick. Outside the space of the vision the material world has disappeared. They mention no humans employed to manage information and fix problems which undoubtedly will appear. All hardware and software comes free. 'Revolution', 'enlightenment' and 'winning' are potent words in business discourse, but ring alarm bells for sceptical larrikins.

This does not describe a sustainable practice, so it could not underpin a revolution. It just uses new technology to support the old dream of Taylorism, that science and knowledge magically provide solutions to the old conflict between capital and labour. The world they do not talk about does not cease to exist. It is essential. This new dark matter contains managers making decisions, and a labour force still doing the work. Linear management is still dominant. It is just not mentioned.

In all these cases, and we could cite many others, 'new paradigm thinking' in business co-exists with the old. The 'paradigm shift' happened a long time ago, but it still has not happened. Paradigm and shift alike are fuzzy. Yet fuzziness is only the beginning of investigation, not the end of explanation. Marxist geographer David Harvey has a usefully nuanced account of the complexity of the present situation in business thinking. He begins by assembling a list, taken from other authors, of the typical contrasts between 'Fordist' (modernist) and Post-Fordist (post-modernist) qualities, similar to what Clarke and Clegg used to define the 'paradigm shift'.

But Harvey does this with a larrikin lack of seriousness: 'the juxtaposition of diverse and seemingly incongruous elements can be fun and occasionally instructive' (1989:338). Out of the game comes an insight into the dynamics of this set of oppositions:

> We then get to see the categories of both modernism and postmodernism as static reifications imposed upon the fluid interpenetration of dynamic oppositions. Within this matrix of internal relations, there is never one fixed configuration, but a swaying back and forth between centralisation and decentralisation, between authority and deconstruction, between hierarchy and anarchy (1989:339).

Harvey's picture corresponds here to what we call the managerialist ideological complex. Harvey does not distinguish between ideology and practice. That is because in the murky reality of Managerialism and management, ideology and practice co-exist inextricably.

3. Chaos and freedom

Globalisation refers to so many things, good, bad and in between, that we need to draw a line and start again, as we try to do justice to this range. Thus far we have looked at globalisation from above, as a vast, often oppressive force. In this section we will look at the more positive aspects. This extract from Bob's journal comes from the same trip we looked at in Chapter 2:

> I arrive at Mexico City Airport on United Airlines flight 832 from Washington. A day ago I was in Brighton, England, visiting my (English-Australian) son and his (Peruvian-Welsh) wife and their new son, my English-Australian-

Peruvian-Welsh grandson. Two weeks before that I was in Australia, stopping off for a day in Singapore. Globalisation is amazing and exhilarating, but exhausting and inconvenient for the ordinary bodies we all still have. I don't understand the interlocking systems that organised my trip so smoothly, and it's better not to calculate the fuel consumed to get me around the globe so fast (bugger those Greenies who invented the idea of a carbon footprint). I just think how strange and wonderful globalisation is, and how much it has shaped my life. (Bob's journal, 15 June, 2007)

Everyone knows that globalisation is important, but it seems too huge and hard to define. It is also familiar, a part of the everyday experience of work and life across the planet. Everyone is an expert on the good and bad sides of globalisation, to a degree, though that does not make it any the less complex and surprising.

Globalisation is a generator of complexity. It weaves many things together, such as Bob's grandson's identities and his own transnational location, an Australian in a US airport en route to Mexico. What is woven together is not always benign. Planes can bring terrorists to their targets, agents of Multi-national Corporations to poor nations to set up sweatshops, as well as activists to many countries to resist big business.

'Globalisation' comes from Latin *globus,* and means 'to make into a globe'. Latin *globus* refers to a dumpling as well as the world. English 'globe', Spanish and Portuguese *globo* all range from small to big: from light bulb, to balloon, to planet. *Globus* is akin to Greek *kolpos,* a fold, bosom or lap. The Greek brings out a quality that seems lost from the modern word, a connection with feminine nurturance. But it is still there, under the surface. Globalisation brings things together harmoniously. It is a hollow which contains everything, a swelling bosom which offers everything. *Globalisation* is a woman. She can appear at the smallest, humblest level, or across the planet. In this root sense, globalisation is positive. We larrikins are pro-globalisation, in this sense.

As we have seen, globalisation is not used like this in business discourse, and the business version of the word has come to dominate everyday discourse. Neo-Liberals tend to label all who oppose them 'anti-globalisation'. In 2000 Neo-Liberal Mexican President Ernesto Zedillo coined the term *globalophobes* (Coronado & Hodge 2004:191-192). This literally means 'fearing the globe', though he meant 'afraid of globalisation'.

This dominant tendency has given 'globalisation' a narrow sense for the left, who often react by using the word as a negative. We believe that this response, however understandable, is counterproductive. Globalisation is a rich word. That is why Neo-Liberals want to appropriate it as their own, and then redefine it accordingly, just as they did with 'Liberalism'. The processes it refers to are so important that Neo-Liberals should not be allowed to steal it. As larrikins we reclaim the name.

For us, 'globalisation' means the creation of any 'globe', any sphere of coherence. Most are smaller than the planet. Some theorists emphasise the role of the 'local' versus the global. Robertson (1993) coined the term 'glocal', a mix of global and local, to capture this dynamic. Our fractal model does not need it. There are many smaller globes at many levels, tendencies to coherence which obstruct larger planetary flows. Nations, cultures, and organisations are all globes, formed by 'globalisation' of a kind. Globalisation is multi-scalar and fractal, formed by similar forces at every level.

Globalisation is also the context for every action of every agent of every organisation. It is part of everyday experience. It is affected by the activities of multinational corporations expanding production and operating across different nations. These activities are driven by the linear plans of CEOs and corporations, but globalisation encompasses much more than them.

Planetary globalisation is so vast and complex that no individual or body has created it, or can control it. Its speed can produce chaos, catastrophe, or order. Globalisation is like a mysterious fluid in which ordinary actions take on strange new characteristics, seeming to obey new laws. In its strange history this fluid has grown and shrunk for reasons of its own. To understand and work with a world like this needs theories which accept and welcome chaos.

Speed is not included in the Latin *globus* but it is essential to the processes by which the planet is today being made a globe. More and more strands are being woven together at ever greater speeds. Bob booked his complicated itinerary in Sydney, and his Australian dollars were rapidly translated into the relevant currencies along the way. Bookings were made and confirmed, and the employees of airlines and airports slightly adjusted their behaviours to accommodate this single journey, along with many other travellers going to different destinations for different purposes.

Planes are less fast but still incredible. Bob's body can go from Sydney, Australia to Mexico City in 20 hours. His jetlag tells him there is a cost, but it seems a small price to pay for a degree and scale of mobility that the planet has never seen before.

This effortless speed seems a miracle, so fast that it can hardly be felt or seen, part of a contradiction which 'globalisation' seems to have resolved between precision and flow. Bob has his ticket and a seat in a specific plane, which will leave from one given destination, and arrive at the destinations listed on the ticket. He does not want creativity here. He is happy to be a single, unambiguous data entry, not the whole complex person he also is. It will allow him to go where he wants, when he wants, within limits.

Yet this simplicity, repeated many times over, produces extreme complexity. A diagram of all individual flights on a given day would look like a giant bowl of noodles flung onto a world map. In spite of this apparent triumph of order (most of the time), precision and flow are still in tension, a balance to be struck afresh in every journey, every system. So are coherence and complexity, order and a kind of chaos. Globalisation is full of contradictions. That is why it is so hard to pin down.

Bob's visit to his son John in Brighton began this reflection on globalisation, so it is fitting that we book-end this euphoric paeon on globalisation with John's sobering comment on the contradictions he saw in it:

> Talking about plane flights in terms of 'bugger the carbon footprint' is deeply unfashionable these days! But also – it misses one crucial point about (economic) globalisation: that it creates a situation where all the resources and consumption gets concentrated into the hands of the affluent – e.g. – you! For you globalisation doesn't mean your village got invaded by US troops because there's the oil which Bob Hodge will need to fly around the world with! To unselfconsciously weave an intellectual web on top of some life-or-death geopolitics pains me.

Scientific models of globalisation

To build up a more complex model of globalisation we begin with the 'Gaia' hypothesis of James Lovelock. Lovelock was an 'independent scientist', sometimes called a 'maverick', with some larrikin qualities. He produced his theory outside the scientific establishment, but since he first announced it in 1974 it has won broad acceptance.

The idea itself is applied cybernetics. Lovelock showed that the biosphere, itself an assemblage of cybernetic units and systems, regulates the conditions of life on the planet within a narrow range. Life has not only adapted to its conditions, it has adapted those conditions to itself. The result is what he calls a single super-organism, an extraordinarily complex and highly functional system that emerged by self-organisation. Gaia is a Prigogine dissipative structure, a highly stable open dynamic system formed at the edge of chaos.

We adapt Lovelock's idea as a model for total globalisation. His 'biosphere' includes all life, including humans. In our diagram we extend it to include spheres/systems he did not examine. Globalisation theory often talks of 'spheres' of economics, politics and culture. Economic globalisation is usually seen as the driver, and the information revolution the driver of the driver. We include all these 'spheres' as globes in this differentiated model of postmodern Gaia:

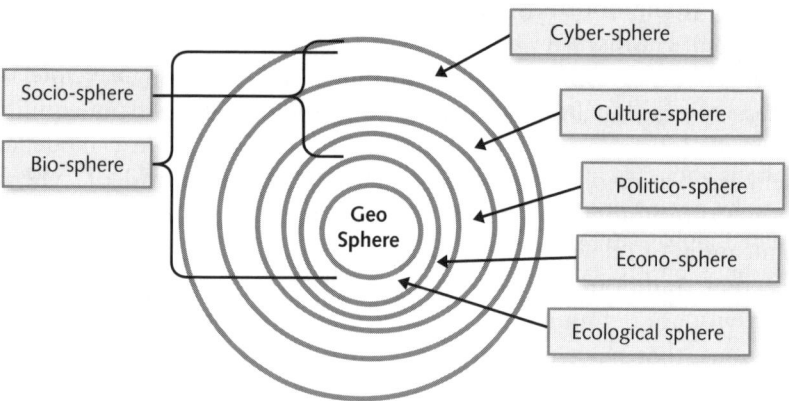

The many globes of globalisering

This simplified diagram of total globalisation allows us to make some important points.

1. Lovelock was concerned with the relationship between only two spheres, the biosphere and the climate system, in which the smaller biosphere stabilises runaway feedback processes that otherwise would have destroyed all life. He did not factor in our other spheres because humans have not existed for most of the planet's history. But now we exist with a vengeance. Climate change science tells us that human activities are an accelerating feedback loop that counters the previous cybernetics of the biosphere.

2. The lines in the diagram make the spheres seem separate, but this is misleading. These spheres are mingled throughout human activity. Lines should be fuzzy and mobile, to capture the dynamic complexity of the global system of systems.

3. A revolution (major change) in any one sphere (e.g. the econo-sphere or cyber-sphere) will reverberate throughout all globes of the system, but not predictably. Cybernetic linkages can magnify or dampen effects. Forces and changes must be studied across adjacent spheres, and not only in themselves. No 'revolution' is self-contained.

Many social scientists who study globalisation use the term 'postmodern' to capture their sense of today's extreme complexity. Harvey (1989:284) identified 'space-time compression' as its distinguishing feature. Speed is crucial. It has the effect of shrinking space and time. What was distant becomes close, and the past or the future come into the present.

We rework Harvey's proposition by framing it within classic non-linear science. 'Space-time compression' echoes Einstein's famous proposition that time is a fourth dimension, because it co-varies with space. Heisenberg's famous 'Principle of Uncertainty' makes a similar point to Einstein, and allows us to take Harvey's idea further. Heisenberg (1989) demonstrated that in the chaotic world of quantum particles, it is impossible to be equally precise in describing both position in space and momentum (changes of position in time), because they co-vary.

Most scientists restrict the application of Heisenberg's principle to the quantum world, but Prigogine argues it applies in all far-from-equilibrium conditions. Likewise, Zadeh's Principle of Incompatibility, modelled on Heisenberg's principle, asserts that in the everyday world of cybernetics there is a threshold above which precision (exact position) is incompatible with meaning or relevance (behaviour).

Different traditions of globalisation studies agree that speed and hyper-connectivity are defining qualities. Space-time compression says more than the obvious, that space is covered more quickly, or that travel time becomes less. Time and space shrink and merge. They co-vary.

At the beginning of the 20th century, Max Weber (1947) used 18th century American Benjamin Franklin to exemplify the capitalist mind, especially as captured in his famous phrase, 'time is money'. We see this as time-money compression, positioned by Weber as the foundational premise of capitalism. This merges with space-time compression to create a three-way equivalence between time, space and money. Money has become the 5th dimension in a post-Einsteinian space-time continuum.

This is not just a recent condition of the postmodern world. It is intrinsically connected with globalisation, whose accelerated forms make its processes more visible. It is associated with capitalism, but not with its conscious aims and logic. On the contrary, it undermines the linear rationality of capitalism, making chaos theories a better guide to the world that capitalism has helped to make.

The globalisation of resistance

Many opponents of Neo-Liberalism describe themselves as 'anti-globalisation'. We argued earlier that *globalisation* is the wrong name for their target. Globalisation is a vaster, more complex context containing currents and counter-currents. The systems that support Neo-Liberalism need contrary forces of larrikin mediation and resistance, without which they are not sustainable.

Many organisations and mechanisms perform this function, usually small and poorly-resourced in comparison to the giants of business and government. They are so dispersed that they cannot easily be counted or organised. The same is true of the biosphere in Lovelock's Gaia, yet both are vital to the health of the planet.

We will not attempt a comprehensive account of these complementary processes. As an illustration, we look briefly at the World Social Forums (WSF), a series of meetings which began at Porto Alegre, Brazil in 2001.

These were set up to oppose the World Economic Forum, held in Davos, Switzerland, comprising major political and economic players in world capitalism.

Responding to criticisms that this opposition to globalisation was itself a global movement, the Porto Alegre group described itself as 'Alter-globalisation'. This change of names allowed the group to sidestep a barrier to thinking about issues of the Neo-Liberal political and economic order.

Contradictions do not exist just between these two forums. In Davos, world poverty is regularly addressed as a core problem for the global capitalist system. Poverty and injustice are key problems for its ideological complex, generating much rhetoric and some action in attempts to counter them. These political institutions realise that they need to be seen to be doing something about poverty, with no real intention of acting on this. Likewise, WSF meetings were supported from the outset by world bodies like UNESCO. Speakers included Noam Chomsky in 2003, trenchant critic of Neo-Liberalism, and Joseph Stiglitz in 2004, Nobel Prize-winner and former World Bank official.

Bob's son John commented on some of the tensions behind this kind of forum from an activist point of view:

> Activists resent the presence of NGOs at 'counter summits' like the WSF not because of 'ideological purity', but because the NGO's represent both organisational structures and agendas which are mostly 180 degrees away from the bulk of the groups involved. The post Seattle movement(s) – as spearheaded by frameworks like Peoples Global Action (PGA) – were about creating a 'network of networks' with indigenous groups, ecological and social campaigns and other counter-struggles all working in a cooperative, non-hierarchical (non)structure, but always working from grassroots upwards. NGO's are seen to insert themselves onto this worldwide movement by pushing for a hierarchy which they are the leaders of, hoovering up the majority of the publicity, and most crucially pushing for a much more reformist agenda than any of the activists want.

But no action can be free of contradiction. The crucial question is how the contradictions are managed, in far-from-equilibrium conditions where linear cause-effect does not always apply.

The numbers attending this forum went from 12,000 people in 2001 to 155,000 in 2005. This exponential growth signals far-from-equilibrium conditions. By 2005 it had crystallised a position around a 12-point program, the Porto Alegre Manifesto. This had 12 points, compared to Williamson's 10 in the Washington Consensus.

Most points were answers to the Washington Consensus. But the final two matched nothing in the Washington Consensus. They dealt with 'democracy'. This was a favourite word for ideologues of Neo-Liberalism like

Friedman and Bush, but did not appear in the Washington Consensus conditions. Indeed, many favoured recipients were dubious dictators or corrupt demagogues.

Yet the Porto Alegre Manifesto does not focus on political systems. 'Democracy' is about the conditions of media that support democracy. These include ending the concentration of media ownership; protecting journalists' autonomy; and supporting alternative media. Non-coincidentally, these are all at home in the informal spaces inhabited by the Larrikin Principle.

Alternative media as David vs Goliath

The Biblical myth of David and Goliath is a useful resource of non-linear reflections on resistance. Tsoukas (1999) used the metaphor to analyse the surprising defeat of an MNC by a coalition of environmentalists led by Greenpeace. Hodge and Coronado (2002) used it to explain the role of the Net for the Zapatistas in Mexico. In 2001 a small group of Zapatistas made a perilous journey, from the jungles of Chiapas to deliver their message to the Mexican parliament. Armed helicopters hovered threateningly overhead. The Mexican army could have exterminated them at any time, yet they survived, thanks to the gaze of the world channelled by the Net.

We will finish this discussion with a more ordinary example of alternative media and the David-effect. *Schnews* is a one-page newsletter produced every week from a small office in Brighton, UK. Bob has known it well for many years, because his son, John, lives in Brighton and has been involved in *Schnews*.

It would be easy to ignore Schnews. Goliath did not take David seriously either. Schnews packs its stories onto two sides of a single sheet. Its-readership is about 30,000 per week, a fraction of the circulation of major British tabloids. How could it hope to have any effect?

Part of the answer, as with the Zapatistas, lies in the role of cyberspace. The myth of David has a similar structure to the Death Star corollary we discussed in Chapter 3. Goliath's heavy body-armour had to have a chink, through which he looked out, through which David's sling-shot could enter. The site of vulnerability lies in the system itself, immersed in a complexity it cannot control.

Schnews is not just the two dense pages it produces every week. It also has a web-presence, with many more hits than copies distributed. Each issue includes web addresses for those interested in following up particular stories. A book commemorating its first 10 years (Schnews 2004) included a 'yellow pages' directory to 800 other sites. The power of any node is a function of its connectedness.

It is not possible to attribute causal effects to any individual link, such as Schnews. It is the network which has its effects, which are non-linear but substantial. We illustrate this with one example. Schnews was only one of many nodes in the network that sustained the worldwide protest movement against the Iraq war. On a single day, 15 February 2003, over 20 million people took to the streets to protest. It is a massive effect of this network.

In spite of the scale of the protest, many asked whether in practice it had any real effects. The Schnews team themselves wondered:

> We went on the marches like everyone else, but (like most people) we didn't hold out much hope that marching by itself was going to be effective. Most people involved with Schnews thought that the only real hope lay in constant and massive acts of disobedience... We didn't stop the war from happening, but that was always going to be a tall order. (2004:247)

Six years later we add another interpretation to the mix. The war was not stopped, but its main architects, Bush, Blair and Australia's Howard among them, are out of office, their credibility damaged. The marches and other activities cannot be shown to have had measurable effects, but it seems evident they had some, in a complex, non-linear chain, over a longer timeframe.

John commented on this argument:

> The measurable effects of the political movements Schnews talks about are often outside the mainstream media – or are misrepresented by it – that's why Schnews, Indymedia, Corporate Watch, and many others bother. Two things have come out of the anti-war protests 2001–2004. One is that the broader population starts to think in the direction of being anti-war, but doesn't do anything, yet – whatever effect that has – because voting in Obama or Rudd hardly shifts the political landscape. The other is that the anti-war movement starts to broaden out and attack the war machine on various fronts (having realised that marching didn't work) – including the anti-arms trade campaigns, climate change groups like Rising Tide (because these wars are directly tied into oil and other resource control), refugee campaigns like No Borders (because the vast majority of refugees come from war-zones, currently headed by Afghanistan, Iraq, Pakistan, and Darfur), and others campaigning to stop Guantanamo/Rendition/Torture. The anti-war movement did 'politicise a generation', but of course marching, and being 'anti-war' are just starting points for actual activities which are anti-war.

Schnews has a particular location and focus. It is produced in Britain, in Brighton. This gives it its own fractal strategy for relating 'local' and 'global'. To illustrate, Issue 308, June 15, 2001, reported on an incident in Brighton:

> Brighton's refuse workers went to work this Monday to find that SITA – the French multinational with the contract to clean Brighton's rubbish – had imposed increased workloads that the workers knew would be impossible to deliver. (2004:295)

This story illustrates Neo-Liberalism (privatisation, outsourcing) and Managerialism (a 'restructure' to cut costs by unilaterally increasing workloads), but these terms are not used. In this case the strike finished well for the workers and badly for the company. These strikers mobilised support from others in the community that was sufficient to pressure Brighton Council into withdrawing its support for the company. This David-and-Goliath structure is found in many *Schnews* stories, no doubt to inspire their core constituencies, the radical and disaffected. Yet *Schnews* also aims at a wider audience. As John said, in an interview for the edited book:

> The good thing about *Schnews* is that it does not compromise content, but yet manages to reach a wide audience – even though it gets called 'populist' by people whose worthy little papers reach tiny numbers. (2004:279)

This scope is reflected in its larrikin style of humour. It is critical, irreverent, disrespectful of all authority, including icons of the left. A regular section, 'Crap arrests of the week', usually shows police looking ridiculous, but the book includes a jokey glossary of terms with satirical content.

'Summit Hoppers' are defined as 'People who use mobile phones, email and cheap air travel to fight... er globalisation' (2004:8). These are activists at the heart of the *Schnews* community. Less jokey is an entry that captures a contradiction at the heart of radical alternative forms of organisation in groups like *Schnews*:

> Vertical non-hierarchies
> Hierarchies and divisions of labour which unavoidably occur in non-hierarchical groups due to people's different skills and abilities, but which can never be acknowledged (2004:8)

Just as Managerialism has embedded contradictions, so too do radical alternative organisations. As John told us,

> This is a perennial issue for campaign groups: see *The Tyranny of Structurelessness*, the famous 1970 paper by feminist Jo Freeman. That does not mean there is no difference'.

The fact that they can be ranged along a continuum only makes differences more significant. Mainstream organisation studies could learn much from the amazingly efficient organisation of the *Schnews* office, and vice versa, without collapsing either into a single, one-size-fits-all model. John describes the *Schnews* methods which do so much with limited resources:

> What you have in mind at the beginning, and what you end up with at the end are two totally different things – you have to be flexible.

The books are an experiment in using all yer right-on decentralised non-linear ways to do things which might otherwise be done very rigidly in the structure of a publishing house. (2004:278)

John uses terms from the managerialist lexicon like 'flexibility' and 'decentralised', but does not use them as rhetorical justifications for linear, profit-driven actions. Yet the expertise of commercial publishers is not disregarded. They show 'things to do', which still have to be done, but differently, for different motives.

We call these practices larrikin-like, aware that most activists and alternative media practitioners have never heard of larrikins. Our aim is not to co-opt them into our term. The Larrikin Principle is just our way to point out that many alternative media and social activists have a remarkable impact. They play a vital role in the improved management of the planet, in spite of their small numbers and limited resources. Their opposition to Managerialism and Neo-Liberalism crystallizes what many dissenters and larrikins in organisations think. The effects they have or channel are non-linear, but in a non-linear world they are just as potent for that.

We finish with John's comments on this section:

> My final thoughts about this section: the most interesting point – which remains unexplored fully here – is the question of non-hierarchies vs. Managerialism. Most campaign groups and groups like Schnews aim to work as horizontal non-hierarchies, and in fact see solutions to the inequalities of the world in the form of (horizontal-leaning) decentralisation and direct democracy, e.g. regional, indigenous and other communities regaining control from centralised governments and MNCs, with an emphasis on taking the democratic process to as grassroots and proactive a level as possible, and that these groups network with other adjacent/international groups for mutual benefit and cooperation (of course this comes after all the despots and oligarchs have been lined up along a wall and shot!). A critique of neo-liberal Managerialism is easy from this perspective, but as attractive as the other model is, it is fraught – hence The Tyranny of Structurelessness. The non-hierarchy is effectively a larrikinocracy, but does anyone actually want that? I guess the answer is a benevolent non-capitalist structure, geared to empowering minorities and preventing greedy or despotic behaviour.

The future of Neo-Liberalism

Neo-Liberalism is somewhat damaged as the international brand-name of Managerialism. In time the term may disappear. But if so it will be replaced by an Ideological Complex with a similar form, another mirror image of Managerialism. Allied with globalisation it has been a vector and catalyst for Managerialism's recent and devastating impact on the planet. Yet a main task of larrikin criticism is to insist on the differences between globalisation and Neo-Liberalism. Globalisation is a mystery demanding new and old forms of science to explain it. Neo-Liberalism mystifies it and makes it harder to explain. But this is hardly surprising. Mystification is what it was designed to do.

Chapter 5

Corruption

'Corruption' is as complex and difficult as our other key terms; larrikin, Managerialism and Neo-Liberalism, but in different ways and with different effects. Where the other words have ambivalence built into them, 'corruption' is usually presented as if it were morally simple and pure. All corruption is bad, by definition.

However, this apparent simplicity is a delusion when applied to complex situations. The simplicity is used as a weapon. To label someone 'corrupt' (e.g. a larrikin) means that nothing more needs to be said. Anyone who tries to defend the behaviour seems to be defending the indefensible. Since it is used so often as a weapon, we need to see who uses it like that. There is a further trick. Accusing someone else of 'corruption' implies you are uncorrupted yourself.

In this chapter we show how to negotiate this tricky terrain, as it affects an understanding of Neo-Liberalism and the Larrikin Principle. If we only show people how to think and talk about the connections and issues we will have done a useful job. Discourses about corruption may be simple in principle but are a minefield in practice.

The final difficulty with 'corruption' is that it remains a crucial theme. Although we question many particular uses of it in business, politics and the media, 'real' corruption is as pernicious as people claim. Our goal in this chapter is to deconstruct illicit uses of the term in order to grasp important qualities of the world system, past and present.

'Corruption' comes from Latin *'con-rumpere'*. *Rumpere* on its own means to break some definite object. *Con*, 'with', subtly changes the word. It makes the process diffuse. An individual crime breaks a rule, but the system of law can remain intact, to judge the act and those who commit it. In ancient Rome *corruption* meant activities which not only broke laws but also eroded the system. This distinction still survives today in English, as in Spanish, Portuguese and other European Romance languages.

It may sometimes be difficult to apply when the system of law itself is dysfunctional or lacks legitimacy. This is always the case, to some degree, especially in turbulent, far-from-equilibrium conditions. This fuzzy area, larger or smaller, around corruption impinges on aspects of the Larrikin Principle. Corruption involves bending or breaking rules. So does larrikinism. However, not all rules are the same. Some are more formal than others; some are written while others are just in the minds of their users, like phantoms guiding organisational practices. Some are laws more than rules.

Corruption has always had close links with power. In ancient Rome as in colonial Brazil and Mexico, Australia and the USA, and in business and political organisations today, 'corruption' comes from within a system. It is normally an unacknowledged part of the system. It corrects dysfunctions and destroys the system. This is where 'corruption' overlaps with the Larrikin Principle. Sometimes it saves the system from paralysis, like informal systems run using the Larrikin Principle. Sometimes it is malevolent and destructive. Sometimes it is both, in a mixture which is hard to separate out.

1. Fuzzy corruption and the fuzzy larrikin

We begin this chapter by looking at situations where it can be difficult to distinguish between larrikins and criminals. Our first example is Paul Hogan, a creative artist and successful businessman, who made millions of dollars creating a form and brand of larrikinism for global circulation.

His *Crocodile Dundee* films were popular domestically and internationally. They took the archetypal larrikin figure, portrayed in Mick 'Crocodile' Dundee, and skillfully but unobtrusively updated it for a contemporary world. Dundee is introduced as a rural character in a rural setting, but he has the street smarts of an urban larrikin. The stories took him between Australia and the USA, and he acquired a US wife, played by Linda Kowalsky, whom Hogan later married in real life. Dundee was a mixture, Anglo-Irish with special affinities with Aboriginal Australians and US urban blacks.

In 2006, an Australian newspaper, the *Sydney Morning Herald*, published an article reporting that Hogan and others were being investigated by the Australian Tax Office for 'serious tax fraud'. Hogan responded in character as Mick Dundee, iconic larrikin:

> To whom it may concern,
> According to *Sydney Morning Herald* on August 24, 2006, I am in so much trouble with the ATO (notice how I just use their initials to protect their identity) that I have hired a 'tax fraud specialist', a Washington lawyer named Scott Michael.
> You got me. Almost. The last problem I had with the ATO was in 1972 when they claimed that I had 'fudged' the overheads on my earnings from my pub "chook raffles". I did hire a lawyer then, my cousin "Shifty Joe" Hogan. He lost the case. He was disbarred in 1973. Since then my relationship with the ATO has been excellent. Probably because they were, by as far as you can throw your hat, the biggest beneficiaries of my movie endeavours...
> I do have a slight problem with "tax fraud specialist" Scott Michael. I had never heard of him. Maybe one of my associates, publican John "the Goanna" Cornell, or "colourful racing identity" accountant "Big Tony"

Stewart hired him. Not only had they never heard of him, he had never heard of us until the *SMH* made the connection.' (*SMH* 14-09-06; 1–2).

Hogan here is Larrikin as Trickster, denying a serious crime by admitting to a joke crime, inflating overheads on a raffle for a "chook" (slang for chicken). He invents nicknames for his friends to make them part of a cast of Mafiosi, joking about a criminal connection which, the joke implies, does not exist.

He plays with the ATO's initials, as though they are part of a game, too, but this joke quietly signals that he is smart and literate. He is precise about real details – the exact date of the *SMH* article, for instance. He has done his sleuthing too. The details about Scott Michael are right, as the *SMH* later admitted.

The letter fuses homespun larrikin Crocodile Dundee with wealthy tax-minimizing (if not worse) business man Paul Hogan, so seamlessly that the joins almost do not show. As we write, we await the results of the long-standing investigation to settle whether Hogan the man crossed the line into crime, or whether he stayed just on the right side, shielded by his larrikin persona.

Ambiguity surrounds larrikin behaviours in Latin American countries, under different names. In Brazil, a key term to indicate larrikin-like behaviour is *jeitinho* (literally 'cutting through'). But some of our interviewees saw only a dark, criminal quality in the term. They report on a negative *jeitinho*, which ceases to be an innocuous strategy to 'get things done' and becomes a more sinister abandonment of ethics for the sake of personal advantage.

Our study coincided with a spate of political corruption scandals in 2004 and 2005 involving the ruling Workers Party in Brazil. Many interviewees felt that this was a massive breach of trust. Below are two typical comments:

> Milena: There are bad *jeitinhos*. In fact, last year there was an explosion of bad *jeitinhos*; one was [a] friend of the other, who was a friend of the other, who was a friend of the other! And what was behind all that? A fortune in money, passed from hand to hand, without income tax; without any propriety. Why? Because of this *jeitinho*. But there's a limit for everything, you know [said emphatically]. People have to know the limits, where they can go without acting incorrectly. And there should be jail. If you've done wrong, you go to jail. But with these people [corrupt politicians], they are above the law.
>
> Daniel: Corruption is happening in Brazil. All these corruption cases, the post-office scandal and so on… Now everything is being exposed, and they still say 'I didn't know; I don't want to know… The reality is that there's been millions and millions involved and no one knows, they say no one knows anything. This is a political problem, I mean… This is where there should be an example of dignity, but all we see is all this *bandalheira* [shabby tricks].

> The government, in my opinion, is like a company, isn't it? But it's difficult to administer it, isn't it? Because there are 'n' points, and the citizens trust Joe Blow as a manager, and sometimes, given certain conditions, he lets the *'vaca ir pro brejo'* [lit. 'the cow fall in the hole'] the way it's happening now.

Milena emphasises 'bad' *jeitinhos* increasing in everyday life at the elite level, while Daniel is more concerned with corruption in political life, but both see deterioration in both spheres. When *jeitinho* covers political or business behaviours, there is less tendency to romanticise it.

Ambivalent larrikins in business

One Australian larrikin interviewee, Nathan, brought out the ambiguity surrounding larrikinism and corruption in business life. He was talking about a senior manager he knew who was involved in the collapse of HIH, a large Australian insurance firm whose fall had similar causes and effects to Enron's:

> And he made it very plain to me that the insurance industry knew what was happening, were well aware that HIH was breaking the rules. Because they were breaking the rules too. And they looked at HIH as it failed and used that opportunity to correct some of their practices before the policemen came to see them. That's if you like another larrikin's behaviour that, it's worse than a larrikin's behaviour, it's a behaviour of someone who is amoral.

Nathan calls himself a larrikin, but he is troubled by the blurred, unstable boundary that sometimes does not separate it sufficiently clearly from criminality. He finally does see a difference, in the morality that larrikins still have, but he is still not sure.

Yet the ambiguity he detects applies to non-larrikins more deeply and systemically. The establishment which is complicit in these practices is not larrikin, in Nathan's judgement, but they blur the edge between criminal and lawful just as much. In so doing they are not criminals, but that makes them only more dangerously corrupt. The system itself produces a silent, deep form of corruption no less dangerous because it escapes detection.

As another example of the fuzzy boundary between criminality and other lawful practices, we use a case that hit the Australian headlines in 2004. A few currency traders for the National Australia Bank lost AUD 385 million in bank funds. They were called 'cowboys', here equivalent to 'larrikins'.

Clearly this was a big amount for a bank to lose. Yet it was not so clear that these men had committed any crime. Currency markets are a far-from-equilibrium field in which small miscalculations can have large consequences.

One Australian management consultant, Louis Coutts, commented on the episode:

> Ever since the early days of the Japanese quality control movement, business has known that if something goes wrong, the cause is usually a poorly designed system rather than malevolent workers...
>
> As companies expand, there is always the danger that their managers can become distant from the daily activities of employees and too concerned with meeting numeric targets (sometimes realistic, sometimes not). Strangely enough, the greater the controls and the more emphasis given to what is believed to be quality assurance methods, and the more incentive-driven a company or division becomes, the greater likelihood of an informal organization developing that is resistant to the dictates of management. (2004:61)

Coutts recognizes some of the paradoxes at the centre of our book. Faults mostly come from the system, he says, generating informal systems, which then become problems. The contradiction is close to the surface: there are and should be informal systems, yet they should not exist. The problem stemmed from loss of control, produced by controls that were too great, too linear.

Coutts deals with a specific kind of organization, located firmly in the developed world. Significantly he does not speak of 'corruption', but that is what he is alluding to. We will use his analysis as the basis for some propositions about connections between corruption and larrikinism, as they reflect and react to core problems of Managerialism and Neo-Liberalism.

1. Informal organizations within larger organizations may develop from people who are committed to the organization and who are dissatisfied with the dominant systems and their bosses. In this respect they embody the Larrikin Principle.
2. These informal organizations can develop in ways that jeopardize the official system. At this point they may be judged to 'corrupt' the official system, since they challenge some rules, and the legitimacy of the rulers.
3. Yet the primary aim of these 'larrikins' may be to serve the organization better by following different rules, defying management edicts. In this respect they preserve the organization from its own official systems. Far from corrupting the official system, they may mean to make it work better, and sometimes do.
4. In this situation, it is often dysfunctional, crisp, linear systems that produce both the problems and the informal practices that attempt to solve them. They can also produce senior management's ignorance of the defects of the dominant system. The root of dysfunction and 'corruption' may be inadequate, linear systems, even when linear bosses intend the opposite.
5. Even when the official system recognizes the value of these informal systems, and relies on them in practice, in official discourse they may still repudiate them.

6. Whatever the original motives and responses, informal organizations may drift into increasingly dubious practices, immoral or illegal. NAB's rogue traders may not have crossed this boundary, but they were already in dangerous territory. Informal organizations inhabit a zone where activities can slide unseen into deep corruption.

2. Pseudo-corruption and the Neo-Liberal complex

Neo-Liberalism has a complex relationship with the theme of corruption. On the one hand, its rhetoric condemns corruption in all its forms, and represents Neo-Liberalism as the best antidote. But as we noted in Williamson's comments on the Washington Consensus, corruption did not in fact figure highly amongst its conditions.

Nor is it inspected as rigorously in developed countries (including Australia and the USA) as in developing countries like Brazil and Mexico. James Wolfensohn, president of the World Bank, referred in 1996 to the 'cancer of corruption' (Seligson 2002:410) but the target is developing nations. Another common phrase, 'culture of corruption', sees it as all-pervasive and systemic in nations like Mexico and Brazil, in ordinary conversations and even in official documents (Covarrubias 2003).

This assumption from above is sometimes questioned in our Mexican data. According to José:

> There is one thing that yes, it is cultural... And it has much to do with violation of norms. And it is national. And I tell you because I see it everywhere and you see it in the results that make the municipality function... In Mexico the issue of norm violations has much to do with the fact that the people are convinced of the idea that it is not possible to do things in the right way.'

Here, José does not praise corruption, but sees it as something forced onto decent citizens by a dysfunctional system.

Yet, whatever corruption is, it plays almost no part in mainstream management education. We examined 5 typical management textbooks from our corpus to see what they said about corruption, and came up with meagre results. Only two had a section on corruption. We extended our search, and found it is often dealt with under the heading 'whistleblower'. All texts had something to say about whistleblowers, but never much. We summarise the results in the following table:

Book	Corruption	Whistleblowing	Total corruption
Davidson	None (0%)	3 pp. (0.4%)	3 pp. (0.4%)
Robbins	2 pp. (0.25%)	3 pp. (0.37%)	5 pp. (0.61%)
Hill	3 pp. (0.6%)	0.5 pp. (0.1%)	3.5 pp. (0.7%)
Waddell	0	3 pp. (0.57%)	3 pp. (0.57%)
Hitt	0	2 pp. (0.3%)	2 pp. (0.3%)
Totals	5 pp. (0.15%)	11.5 pp. (0.35%)	16 pp. (0.52%)

The figures make a stark point. Only 40% had anything on corruption. All had less than 0.5% on the theme. Whistleblowing is covered better, but still less than 0.5% on average. As far as mainstream management is concerned, corruption hardly exists as a problem for managers in the developed world. So if it does. they will not be equipped to understand or deal with it.

We described whistleblowers like Jeffrey Wigand as larrikins, but whistleblowing is mostly included under 'ethics'. Whistleblowers will need ethics in pathological abundance, as the literature shows. The need for more ethics in business is a mantra of business education today, to solve problems exposed by the activities of Andersons and Enron in the USA, and HIH in Australia.

We do not argue against ethics. However, in this context it makes corrective measures seem optional, what businesses will do themselves, out of the goodness of their hearts. It supports the Neo-Liberal premise, that business does not need regulation. The market can be trusted to do all that is necessary.

Nor do these management textbooks consider that there might be anything dysfunctional about organisations which need whistleblowers. As we noted in discussing *The Insider* these are a marginalised sub-category of the Larrikin Principle. The fact they are so marginalised makes them like canaries in mines: indicators of something seriously wrong in organisations or industries, and in the dominant form of management that makes them so necessary and so difficult.

Wars against corruption

The preferred Neo-Liberal means of ensuring higher standards against corruption is to provide better levels of information. This is done primarily under the auspices of an international corruption-monitoring agency, Transparency International (TI). This aims to measure 'the extent of corruption' in nations, defined as 'the misuse of public power for private benefit' (Lambsdorff 2004:3).

This strategy assumes that the only reason corruption continues is because some or all parties are unaware of its existence or extent. We suggest this is not a reasonable assumption. Nor are the terms of the definition well-designed to meet or measure the real scope of 'corruption' as we have defined it, as a systemic threat to a whole system of society and governance. This definition foregrounds public power as the main site of corruption, as in Neo-Liberal ideology, which limits corruption to the sphere of government behaviour. If the misuse of power for private benefit comes from private resources, does it mean it is not corruption?

We begin our encounter with corruption in Mexico, a nation with a high corruption index according to *Transparency International* (2008). Closer to the ground, where culture is lived in everyday life, anti-corruption campaigns can look different. Here is an extract from Gabriela's notes of a field trip to Mexico in 2006:

> When I was driving in Mexico City in one of the wealthiest neighbourhoods, Polanco, a street billboard caught my attention. At first I did not realise that it was part of an anti-corruption campaign by business organisations. I passed in front of it many times, and only when I was able to read all parts of the text and image did I appreciate its relevance for my research. Then I stopped to take a picture. (Field notes, May 2006 in Coronado 2008)

Gabriela was initially puzzled by this billboard (image below) and the phrase which dominates it. *'Pa'l chesco'* is a popular phrase, a slang form of *'Para el refresco'* (For a cool drink). The slang identifies this as the language of people of low income who provide a service. The expression usually comes as a polite request: *Señito, ahi pa'l chesco no?* (Dear Ma'am, something for a cool drink, no?). Literally it asks for a tip to buy a refreshing drink after working hard under the sun. Gabriela has heard it all her life, from people who serve petrol in garages and receive no salary, from low-paid workers like *'macheteros'*, who carry bags of cement or bricks to private houses, and the employees of big companies who make home deliveries. She did not see similar campaigns in other less affluent neighbourhoods. She wondered: why run it in this one?

At first glance we might suppose that the people running this campaign share a common language and culture with the ordinary working people who use this phrase. But the phrase in the next line *tenemos sed de honestidad*, 'we are thirsty for honesty' implies that the problem of corruption in Mexico emanates from workers, not from the businesses who pay them so little, and themselves may engage in many other forms of corruption. At the foot of the image the speaker is identified, the Council of Communication Civil Association, who call themselves 'the voice of business'.

Image: Billboard in the streets of Polanco, Mexico City

The two phrases, *Pa'l chesco* and 'thirsty for honesty', polarise the values and language of the two sectors. Business claims it is thirsty for honesty, and it does not use such slang. Common people who are literally thirsty and cannot buy a soft drink are to blame for corruption. Corruption apparently speaks with a lower class accent.

But if asking for a tip is interpreted as corruption, what actions by those 'thirsty for honesty' would say no to it? Impose a new rule forbidding low-payed employees to ask for tips? Or if we believe in the good will of these businesses, 'united against corruption': would they increase wages so that workers do not need to ask the question *'pa'l chesco'*?

Larrikin sceptics may suspect that they will just continue with their campaign as 'socially responsible businesses', saving taxes in the process, and building their moral image. The example shows starkly some common features of anti-corruption discourses.

1. 'Corruption' is typically what others do, even or especially when real doubts may be raised about corruption associated with those who use the term.
2. Accusing these others of corruption is used to imply that the accuser has integrity and is not corrupt. The opposite of this binary logic is equally probable: those accusing others of corruption may also be guilty of it.
3. For this reason the cybernetic function of anti-corruption campaigns may be to shift the blame for corruption within the system, rather than reduce it.
4. Anti-corruption campaigns may have as their main aim and effect to mask and protect corruption, increasing not reducing overall levels of serious corruption.
5. Such campaigns are examples of another form of corruption, not usually labelled as such: corruption of conditions of discourse, whatever gives utterances reliability and sustains truth, emptying out whatever is said of solid content.
6. Thus the larrikin war against 'bullshit' also targets bullshit about 'corruption'.

Reporting third world corruption

Corruption occurring in its expected place, in the developing world, is a staple theme in Australian media. Yet the way it is reported commonly carries a further Neo-Liberal spin. In 2006, for instance, Australian newspapers took notice of a corruption scandal in Brazil that was affecting a Presidential election between President Lula and his main opponent, Geraldo Alckmin. Under the heading 'Corruption scandal tightens Brazil poll', the *Australian* reported:

> Mr da Silva saw his once-commanding lead evaporate on the eve of the vote after his Workers' Party was battered by allegations that party officials tried to pay $US770,000 for a mysterious dossier supposedly containing documents, photos and DVDs apparently linking Sao Paulo gubernatorial candidate Jose Serra to graft when he was health minister between 1998 and 2002...
> A beaming Mr Alckmin made it clear that clean government would be one of his central themes.
> Major newspapers ran front-page photos over the weekend showing piles of money allegedly meant to buy information linking Mr Serra to the ambulance procurement kickback. (*Australian* 2006:8)

Australian readers of this newspaper, one of the Murdoch stable, would see no reason to think twice about this item. Corruption plus Brazil plus left-wing President Lula equals the taken-for-grant known world. Yet the item repays a closer look.

The *Australian* took this item without comment from a wire service, AP. Yet its biases are easy to detect. For instance, President Lula is called 'Mr', stripping him of the title he earned in a previous election. On closer scrutiny, the 'scandal' surrounds a dossier on another politician, Serra, paid for by Lula's party, which 'apparently' links Serra to graft.

The Serra story sounds like real corruption, if it is true. But getting and even paying for documents that show it is not corruption by TI's standards. This sounds more like a counter-attack by the truly corrupt, shooting the messenger. The 'major newspapers' running the story on their front-pages are aiding and abetting the trick with all their resources. That is discursive corruption.

The same item reports 'clean government' would be one of Alckmin's central themes. Clean government by the Brazilian Centre Right? This illustrates our first principle of anti-corruption campaigns: "Corruption is typically what others do, even or especially when real doubts may be raised about corruption associated with those who use the term".

Corruption is indeed a serious problem in Brazil, and a spate of scandals has touched Lula's government. But corruption was and is also endemic in the Centre-Right parties. The tradition goes back to the Generals in the 1960s, with roots in earlier neo-Colonial regimes, which operated in ways similar to Mexico's.

One outstanding example of this in Brazil was President Fernando Collor de Mello. In 1991 he was the first democratically-elected president following the military dictatorship. His platform combined Neo-Liberal economics with anti-corruption rhetoric. He marketed himself in the media as '*o caçador de marajás*' [the hunter of Maharajas, *marajás,* a Brazilian term for corrupt wealth], but he was impeached on corruption charges. His political rights were suspended by the Federal Senate for 8 years (Menezes-Filho & Vasconcellos, 2004).

The biased reporting of the Serra scandal, plus either-or logic, feeds the idea that the developing world is corrupt, and therefore the developed world is not. By attacking Left-Wing politicians it implies that corruption is a property of the Left, that the Right therefore is not corrupt.

Neither implication is justified, but both sustain the Neo-Liberal ideological complex. According to this the developing world is the only site of corruption, yet not corrupt (if it has Neo-Liberal policies in place). In reality corruption is present in both developed and developing nations, even more so in developing nations under Neo-Liberal regimes.

Measuring corruption

'Corruption' plays a significant role in the way international business operates in a Neo-Liberal framework. Supposed levels of corruption are part of the input into assessments of 'country risk' made by businesses deliberating on whether or not to enter a specific country. US management writer Michael Porter talked influentially of the 'competitive advantage' some nations have over others in a single world economy (2001). Business students are taught (see, for instance, Lassere 2003:160) to look at 'corruption indexes' as they imagine doing business with 'host' nations in the developing world, e.g. Mexico or Brazil. Country 'attractiveness' leads to more foreign investment (Kohler et al. 1993). Corruption, it seems, makes countries look unattractive, which is bad for business. Lowering corruption must be good.

The whole process sounds simple, linear and rational as described like this. International bodies such as Transparency International provide measurements of corruption, using a common basis for measurement. The measurements are so precise they distinguish between different countries, and within a country over any given period of time (as reported by country Transparency International branches). So countries move up and down from year to year. Corruption, it seems, can be easily and objectively ascertained, and factored into business decisions.

We include below some data from Transparency International. The Corruption Perception Index (CPI) measures corruption levels in a country, and the Bribe Perception Index (BPI) refers to countries that pay bribes in their export activities. These are mostly the perceptions of business people, with no basis specified for their assertions.

Table: Corruption Measurements

Country	CPI 2007 Rank/100	CPI 2008 Rank/100	BPI 2006 Rank/30	Rank/30
Australia	11	9	3	8
USA	20	18	10	9
Brazil	72	80	23	17
Mexico	72	72	17	20

Source: www.transparency.org

In this table, the developed nations Australia and the USA are consistently in a different category from the Latin American countries Mexico and Brazil, thus confirming Neo-Liberal ideology. However, the USA moves closer to the Latin American countries in its propensity to pay bribes.

If we took these figures as precise indicators of something happening in the real world, we might note that Australia had a serious fall in the BPC between 2006 and 2008. We might explain this by the fact that in 2006, a major bribery scandal exploded in Australia, involving UN sanctions-busting and bribery on a massive scale over wheat sales to Iraq. This would warrant some fall as indicated here. Even so, it points to problems of this measure. The bribery happened before 2006. Only the news came in 2006. This list records perceptions, which may be influenced by media publicity. Bribery that is not perceived will not count.

Brazil and Mexico stay much the same in the CPI between 2007 and 2008, but they move in contrary directions on the BPI between 2006 and 2008. If this movement were treated as significant, it would be interesting to note (as Neo-Liberal commentators did not do) that Brazil under a Centre-Left government improved five places on this measure, while Mexico under a Neo-Liberal government dropped three places. So much for the positive effects of Neo-Liberal regimes on corruption.

We do not make these points, on this data, because we believe the figures are not reliable enough for such fine judgements. That indeed is our main point. We are not against arguments based on numbers. We are against shoddy numbers treated as if they were carefully collected and analysed. That is in itself a kind of corruption.

Transparency International is supplemented by local branches, TM in Mexico, TB in Brazil, which broke away from the international body in 2007. In this discussion we stay in Mexico, with a text collected by Gabriela during her 2006 field trip. A press release of May 9 from TM, *Transparencia Mexicana*, was headlined: 'Mexico, stuck in its corruption levels: Transparency Mexican'. It reported results from the latest application of the National Index of Corruption and Good Governance in Mexico (www.tm.org.mx).

This headline declares that Mexico is 'stuck in' its levels of corruption, which suggests that everything done so far has failed. The reader learns later in the text that the figures do not say this. There have been improvements in some sectors. Actually, 'half the country's federal entities reduced the occurrence of corruption' in some services. The figures paint a more complex situation. The headline simplified this, emphasising only a negative. Why?

We will not examine news items on corruption in Brazil in detail. Here we note only that in general they emphasise political corruption rather than this kind of minor corruption. A possible explanation for the difference is that the government in Mexico is Neo-Liberal, whereas the Brazilian government of Lula is Centre-Left, though it has not broken completely with Neo-Liberal policies.

As we cut deeper into the Mexican data we find similar uncertainties as in the CPI regarding what is being measured, and what the measurements might mean. According to the document the Index registers the payment of bribes, '*mordidas*', declared by households, '*hogares*', literally 'homes'. This tacitly reinforces the idea that corruption comes from everyday culture. By measuring 'homes' that declare they have paid bribes, it implies that these 'homes' are the agents in these acts of corruption. This is the same inverted logic we saw in the *Pa'l chesco* text, which laid the problem of bribery squarely at the feet of those corrupt, sweaty workers who ask for a tip.

The text makes this sense more overt: 'The results published this morning raise an alert about homes' propensity to pay bribes'. This and similar phrases in the document create a weird image of thousands of households driven by some genetic (cancer of corruption) or cultural propensity to pay bribes, perhaps pressing money into the hands of unwilling government agents or virtuous private businesses.

Yet strictly speaking, the figures at best show the honesty of these ordinary Mexicans answering the questionnaire. One trouble for TM in compiling figures on corruption is that the truly corrupt will not fill in forms. Those engaged in hard core corruption, twisting basic processes of government, committing fraud and other serious crimes, are hardly likely to answer TM questionnaires. If they did, they would probably lie. That is the kind of people they are. Their propensity to be corrupt is the core problem, not their victims' vulnerability.

No doubt TM adopts this strategy because it is easy and safe. However, it illustrates the joke about the drunk looking for his car keys under a lamppost. A passer-by gets down on his knees and helps him, to no avail. 'Are you sure you lost them here?' he asks the drunk. 'No,' the drunk replies. 'I lost them over there, in the gutter. But the light's better here.'

It is easier to get responses from ordinary homes than from the truly corrupt. That does not justify blaming these ordinary homes for the corruption. That would be like the drunk blaming the lamplight for losing his keys.

Irma Sandoval, a Mexican academic commentator, made a similar point:

> If citizens in general pay so much money in bribes to facilitate the connection of electricity, to get their cars out of the pound or to enrol their children in schools, the amount of money that big business is ready to offer to clinch juicy contracts with the government must be infinitely more. Unfortunately, TM has not troubled itself to meddle in this sphere. Its timidity is surely related to the fact that its principal sponsors are some of the most powerful in the country. (2008:23)

Mexican larrikin José made a similar point. He queried the fact that Enron was a former contributor to *TM*. José criticised the ideological bias of *TM*'s measurements, which were designed, he said, to 'make a negative judgement on debt levels (of Mexico)'. He contrasted what they measured with what they do not measure. His example was the unfair tax system, under which some big firms pay no tax:

> It is corruption that they don't [pay], and it is corruption that they pay a lawyer to sue the tax office instead of paying. The tax office pays them. That is corruption. It is not in the [TM] list because it is legal. Because they take advantage of some legal loop-holes that they made themselves.

Such problems of the 'Corruption Index', in its International and Mexican versions, illustrate the validity of Zadeh's Principle of Incommensurability. Above a certain threshold, attempts at precision lead to absurdity or irrelevance. The phenomenon of corruption produces and grows in far-from-equilibrium conditions. The figures produced by TM indicate that corruption is rampant in Mexico, and that is the case, as everyone knows. But as soon as the figures, collected in such circumstances, are pushed beyond vague statements, they cease to be relevant or meaningful.

TM blames the wrong people for the wrong actions. Those wanting to function in Mexico, as citizens or business people, would be better off without them. TM measures of corruption are not more comprehensive, deeper or more valid than what is already known. What is worse, they exploit the fatal fascination of Managerialism with quantification to create an aura of legitimacy for their rhetoric.

TM's worthless figures about corruption occupy the place which should be occupied by better figures. That is itself a kind of corruption. Their 'spin' castigates Mexico (ordinary Mexicans) for failing to improve. This leaves the authorities who mount campaigns against corruption looking like well-meaning but helpless victims of corruption that wells from below. The grand corrupters (who may include the same authorities) are untouched, free to continue their ugly ways outside TM's scrutiny. In cybernetic terms, this campaign against corruption may help to increase it. The 'spin' masks the corruption that exists in Mexico, and is itself a kind of corruption.

3. Discursive corruption

'Corruption' usually has a limited scope in business discourse, referring to the 'corrupting' effects of money on relationships of power and responsibility. In this section we take it in a more general sense, as something that affects other important systems. Discourse and money are themselves systems which are liable to corruption in similar ways, and each of these systems interacts, entailing mutual effects on the others. That is the nexus we will briefly explore.

'Spin' is not usually associated with corruption. However, as in the TM case, spin normally accompanies other kinds of corruption, in a multiplier effect that makes corruption harder to detect or resist, more potent in its effects. It is also a form of corruption, of the system of discourse.

'Spin' is associated today with what 'spin doctors' do, professionals whose job it is to twist the truth to benefit their employers. This fits the TI definition of corruption, since a public good, the truth, is twisted to serve a private interest. But 'spin' has a wider, richer set of possibilities. It belongs to a family of words which stem from Indo-European *spa* or *spi*, to stretch.

From this root came 'span', to cover, and 'spin', to weave a web. 'Spider' also comes from this root.

'Spin' as a verb is part of larrikin lore, in the phrase 'spin a yarn', to weave bits of the truth together into a story bigger than the sum of its parts. Yet this co-exists with larrikin scepticism towards the quality. Larrikins enjoy spin as well as see through it. We do not remove this ambiguity from 'spin' or the Larrikin Principle, while maintaining a fuzzy line between better or worse tendencies.

In our interviews, we used the category 'discursive corruption' to describe a common cause of problems within organisations, where systems broke down or did not work well because of the distorted discourse of bosses or managers. Our respondents found it a significant source of problems. In 52% of such stories, spin was the problem or part of it. Sixty-eight percent of reflections mentioned it as a source of problems. Peter commented on the high-sounding rhetoric of his bosses in a dysfunctional organisation:

> The people at the top, they do not, they were only in it for their personal gain, both financial and ego! Then the value of the organisation was really being undermined.

This quality, to these interviewees, does not smooth out problems. It creates them.

The cybernetics of spin

Why and when is 'spin' dangerous, a form of corruption or a positive factor? How does it relate to basic processes of communication in individuals and societies, in organisations at every level? We will draw again on the ideas of Gregory Bateson (1971) which we introduced in Chapter 2. The language and thought patterns of schizophrenics are symptoms of extreme breakdown, total corruption of normal systems of language and thought. Bateson developed a cybernetic model for this corruption of sense itself.

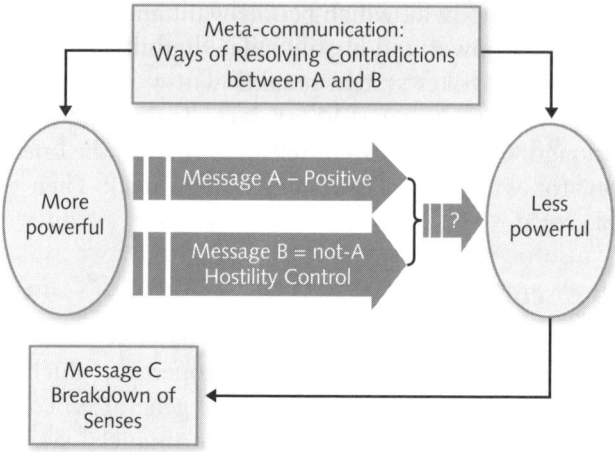

Word salads, confused meanings corrupted system

In ordinary families, Bateson said, adults, the powerful ones in parent-child relations, continually send double messages to children. Message A is positive ('spin') contradicted by message B, a negative message of power and control. Children normally develop a meta-communication loop to sort these contradictions out. They become functional handlers of contradiction, not helpless victims of parental 'spin'. In schizogenic communication the meta-communication loop is damaged. People cannot sort the messages out. Instead of balancing A and B they produce nonsense. They cannot function.

This model applies to the contradictions of Managerialism, in conditions in which 'Theory Y' is found more in rhetoric than practice, but employees cannot challenge the contradictions, to themselves or others. It also applies to contradictions in the Neo-Liberal complex. One such contradiction is the Neo-Liberal attack on regulation itself, in the form of regulations to regulators that they should not regulate. This premise of Neo-Liberalism is literally enough to drive people mad.

This model allows us to make some key points about 'spin' and corruption.

1. 'Spin' as such is not necessarily corruption. So long as there is some way of sorting out the exaggerations and distortions, it does not corrupt the system.
2. 'Corruption' in the proper sense refers not to individual acts (e.g. lies, fraud) but to what happens to the *system*, either by acts repeated so often that they become normalised, thus affecting the system, or because they damage key regulative mechanisms of 'meta-communication'.

3. In this model, the people are held together by affective bonds, of love, respect, loyalty, duty etc., and also by power. Message A typically expresses love or goodwill, which people want and need to hear. Message B often expresses power and hostility. The inability to sort out issues of love and power provokes system breakdown.
4. 'Blaming the victim', as the TM strategy does, is another schizogenic double message. Ordinary citizens are presented as the ones to save from corruption, for whose sake campaigns are waged. Then they are held responsible for it.

Money as a discursive system

Corruption refers to whatever causes systemic damage to a system. Finance and discourse are systems with comparable properties, which we explore in this section, in a speculative inquiry into more general processes of corruption. We begin with the ideas on corruption in financial systems of Milton Friedman, whom we have already met as the champion of Neo-Liberalism.

Friedman targeted hyperinflation, as something that corrupts the economic system itself. He encapsulated his ideas in a doctrine called Monetarism, which proposes that inflation is primarily produced by governments oversupplying money. He called inflation 'a disease, a dangerous and sometimes fatal disease, a disease that if not checked in time can destroy a society' (Friedman & Friedman 1980:298). Wolfensohn said the same of corruption (Seligson 2002:410).

Hyperinflation famously ravaged the German and Russian economies after World War I and the Zimbabwean economy in the early 21st century. Hyperinflation has also been a chronic problem in Brazil, causing many of its financial crises. Friedman's analysis of the problem is persuasive, up to a point. His solution is less convincing. In a simple binary he asserted that the cause of the problems was government failure.

We can see how this half-truth and false logic played out in practice in the case of Brazil (Friedman & Friedman 1984)

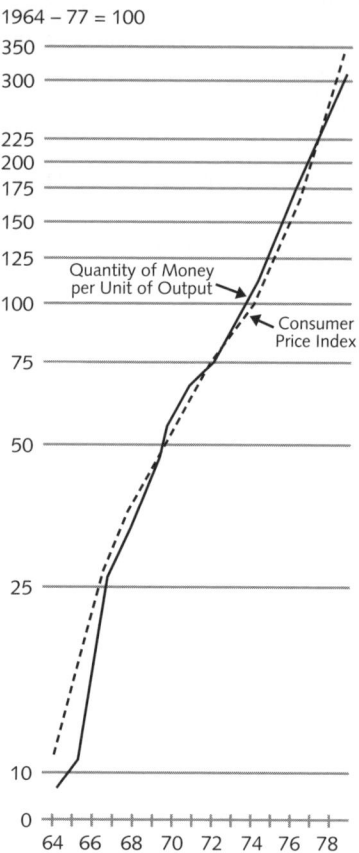

Friedman does not draw attention to one feature of this graph: the vertical scale is logarithmic. That signals a power-law, marking far-from-equilibrium conditions. With this curve, if nothing prevents it, the economy will go beyond a threshold, and collapse. The Brazilian economy did so in 1980, just as Friedman was writing his book. The Brazilian boom collapsed, and the Generals who had run the country up till then decided to try democracy instead.

Brazil's economic reality had layers of complexity not included in Friedman's account, relevant to a more comprehensive theory of corruption. Firstly, the Generals' rhetoric mirrored the exponential growth in prices and money, partly driving it, encouraging unsustainable euphoria in the market. Secondly, the Generals broke with the Washington Consensus on fiscal responsibility, but otherwise were pro-market.

We agree with Friedman that hyperinflation is bad, and corrupts the whole monetary system. Monetarism's analysis is right, up to a point. Marx had already said as much in *Capital,* in 1867. If double the amount of paper money is produced, Marx said, it will be worth half as much (1972:108).

As the current financial crisis reminds us, printing money is not the only way to corrupt financial systems. 'Private' interests are creative in finding other ways of corrupting the system, through various dubious financial instruments. To extend Friedman's argument we use Gresham's Law, which he also referred to appreciatively: 'Bad money drives out good' (1980:296).

Sir Thomas Gresham was financial adviser to three 16th century English monarchs, Protestant Edward V1, Catholic Mary, and then Protestant Elizabeth I. He was a successful capitalist and master of chaos theory who flexibly navigated turbulent times, became a billionaire, and founded the London Stock Exchange.

His law explained hyperinflation. It was driven, he said, by debasing the coinage. 'Bad money', coins with more impurity (cheaper to produce) circulated alongside 'good money', coins made of purer metal. In this situation the authorising government said that each was worth the same amount, so cheaper coinage displaced the more expensive. More and more money circulated, producing runaway inflation.

We see Gresham's law as an effect of discourse. 'Corruption' arises when an authority exercises power over meaning. When governments use their authority to define bad money as (equal to) good, the whole system is corrupted. In this way corruption is the result of an unholy alliance between power and meaning.

This observation echoes another famous saying, by 19th century Catholic English historian Lord Acton: 'All power tends to corrupt, and absolute power corrupts absolutely'. We offer a synthesis that builds on Gresham, Friedman, Acton and Bateson; applying at all fractal levels between global to local, to discourse as well as money:

> *When power interests control meaning it tends to corrupt the system, favouring less worthwhile meanings over more worthwhile. The greater the role of power, the worse the tendency to corrupt.*

This principle applies equally to Managerialism in the micro-world of organisations. In Chapter 3 we read about Tony reporting on a bonus scheme in the Australian bank where he worked. Management assessed eligibility under two columns, one for outcomes, the other for how they were achieved. According to him, management used the second column to avoid paying out on the first:

Interviewer: Wouldn't that create bad morale?
Tony: Ah! Shocking!

Here the legitimating authority, the manager, arbitrarily declares that 'bad money' (doing things her way) is the same as 'good money' (meeting targets). As a result, arbitrary judgements drive out concrete outcomes. The incentive scheme collapses, with the spiralling loss of morale. The parallel with financial crises is close.

Grand corruption in Larrikin land

We begin our investigation into deep corruption in Australia, with a news item about a commission of inquiry into sanction-busting kickbacks to Iraq by a Government agency. The story by journalist David Marr reports evidence given by two top executives from the giant multinational BHP-Billiton to a Commission of Inquiry headed by Justice Cole.

Marr's commentary on Tom Harley, one of these executives, drips with larrikin sarcasm.

> Tom Harley slipped into the witness box like a grey schoolboy taking his seat in class. Neat and willing, he took the oath in an accent best described as Melbourne Oxbridge... (2006:4).

In the larrikin tradition he mocks a representative of British colonial power for being so obviously that. Marr is not a victim of this discursive corruption. He sees through it easily. But in spite of Marr's larrikin wit, Harley is a top manager, and Marr merely a journalist.

On this occasion, Marr mocks the contradictions between Harley's evidence and the evidence of Phil Aiken, his boss, who followed him to the stand. Where Harley maintained 'I always thought it (grain worth AUD 5 million) was intended to be a gift', his boss insisted that it was always a commercial deal. Marr comments:

> So two men from the same company, on the same afternoon, represented by the same lawyers, before the same inquiry, find their reputations in the commercial world now depend on Terence Cole believing that they've been sincerely, absolutely and in complete ignorance at odds with each other over this deal for a decade. (2006:4)

What Marr is mocking is a serious matter, which encapsulates many symptomatic issues arising from the case. These two top executives had only to keep denying their knowledge and complicity to get away with it, even if they contradicted each other.

Paul Volcker, former Chair of the US Federal Reserve, had investigated claims that various agencies were systematically scamming UN Oil-for-Food policies that allowed humanitarian aid to Iraq, in spite of the sanctions imposed on the regime of Saddam Hussein. He found the No. 1 scammer was the Australian Wheat Board (AWB), a government-managed monopoly-based instrumentality for selling Australian wheat. Volcker estimated that AWB laid out USD 221.7 million, four times more than the next worst offender, a Thai company (Bartos 2006:20).

This should have been seen as very serious corruption. Australia is a medium-sized developed economy, so it should not rank first for sanction-busting kick-backs. Yet media coverage was surprisingly mild. Journalists normally called it a 'scandal', not 'corruption'. One effective way in which Australians maintain the belief that they are not corrupt is avoiding the word.

The government was forced to do something. It set up a commission of inquiry, but was careful to frame the terms of reference tightly, so that explosive issues fell outside Commissioner Cole's brief. The hottest issue was the degree of knowledge, or involvement, of the government, including Prime Minister Howard, Foreign Minister Downer, and Trade Minister Vaile. There were pleas of extraordinary ignorance and poor memory, e.g. Tom Harley and Phil Aiken. The ploy was successful, because they had appointed the judge and defined his terms of reference.

Bribery on this scale is criminal, not just unethical, under Australian law. But the way the matter was dealt with made it more deeply corrupting. Deep corruption included Government actions to maintain deniability by framing the Commission brief so that it could not easily arrive at findings against it, while claiming that the Commission was a full and open inquiry.

Corruption continued with the non-evidence given by the key players, and its acceptance by the Commissioner. Only a minority of journalists like Marr exposed what was going on. The system exercised its power to control meanings. They said, with the full authority of the law, that black (i.e. government and big business complicity in bribery and sanction-busting) was really just a harmless shade of grey.

Deep corruption, Mexican style

We have queried what Corruption Indices really measure, and emphasised that much that is labelled and condemned as 'corruption' is better seen as creative informal practices, necessary for survival. We point out the ideological bias that makes Australians (and Americans and other developed nations) believe that they are by definition free from corruption. All that is true, as far as it goes. 'Corruption' is highly complex, in both the developed and the developing worlds. Yet in the case of Mexico and Brazil, corruption profoundly damages these particular national systems, as all reflective Mexicans and Brazilians know only too well. Corruption there exists on a scale that dwarfs what happens in Australia, Europe, or America.

To illustrate, we take a case that arose in Mexico in 2006. While the Australian government was mobilising its resources to minimize the AWB issue, the Mexican government was dealing with a recurring, chronic problem with Pemex, the state-owned petroleum company. Official reports revealed that this was a badly-run organisation, with massive 'irregularities' – a euphemism for corruption. By 2008, the Neo-Liberal government was doing its best to push through a form of privatisation. Privatization is a policy plank of Neo-Liberalism, whose ideology says that privately-owned entities will be more efficient, and less corrupt.

We will look closer at this case, because of the complex ways it focuses the issues. Pemex is a state-owned operation which is hugely inefficient and deeply corrupt. A few examples will be sufficient to show the problem. For 2006, the final year of the government of President Fox, government auditors discovered 'severe anomalies' in many of the accounts. One of these benefited a company, Oceanografía, which was run by the two sons of Marta Sahagún, second wife of President Fox. The auditor noted:

> The contract was awarded to this company – when there were other bids with better offers – for 198,000,000 pesos, $US18 million, of which 133 million pesos were paid in 2006, and work was acquitted to this company without reports being submitted that form part of the necessary paper-work. (Ballinas and Gardunyo 2008:8)

USD 18 million is not *corrupción pequeña* (small corruption). It is more than a small bribe or petty cash.

Pemex is not a well-run public instrumentality. The accounts of the Federal Treasury in 2008 stated that Pemex had achieved record earnings in 2007, but 89.7% of its investments in Exploration and Production came from foreign debt (Rodríguez 2008:18). Our business-literate readers will see this as poor management. Yet it makes sense from another point of view, as one Mexican commentator noted:

> The most recent federal administrations, in agreement with Neo-Liberal circles, have claimed to minimize and even eliminate State intervention in strategic sectors of the country, such as energy. The indebtedness of the state-owned corporations and their growing financial restrictions are very convenient measures to detract from their capabilities, and show that they are poor investments which impose a heavy burden on Mexicans, so as to justify in this way their privatization, total or partial. (*La Jornada* 2008:2)

This way of running things is not simple incompetence. On the contrary, the government is able to fulfil many aims simultaneously. Through contracts offered to multinationals, they gain many of the benefits of privatization without the inconvenience of competition. For instance, in 2004, Halliburton, whose former President, Dick Cheney, was currently Vice-President of the USA, received USD 1,221 million, over 25% of the total contracts for the year, from subsidiaries of Pemex, its privatised arm (Rodriguez 2005).

Pemex 'irregularities' created massive funds to use for various corrupt purposes, including campaign funds for the ruling party. At the same time they discredited the idea of state-owned enterprises, justifying a sale which will enhance their Neo-Liberal credentials. This in turn allowed a feeding frenzy for those managing the privatisation, and their business mates. It is a win-win-win policy, in which the main losers are the Mexican people.

We have discussed this typical-enough case at this length because it illustrates so neatly the complex relationship between privatization and corruption in a nation within the Neo-Liberal sphere of influence, like Mexico and other Latin American countries. State-run instrumentalities are utterly corrupt and inefficient, as Neo-Liberals like Friedman claim, but big business played a major role he would ignore.

The Mexican government is highly corrupt, and it espouses Neo-Liberalism. Yet this corruption is not a product of Neo-Liberalism, imposed on a previously clean state. Previous Mexican governments, including an unbroken 70-year period of rule by a single party, the PRI, laid down the main patterns of corruption, which its successors took over and used for their own benefit.

There is a similar situation in Brazil. The conditions for the emergence and continuous practice of political corruption were in place long before Neo-Liberal governments. In both countries this problem stems from a long line of autonomous power elites drawn from conservative sectors of society, who make decisions according to their own interest. These interests are accommodated within the contradictions of Neo-Liberalism across its national and global forms.

The consequences of corruption

We will stay in Mexico for the moment, to trace some ramifications of this kind of corruption. It is now accepted that in 1988 the Neo-Liberal presidential candidate, Salinas, from the official party, the PRI, used fraud to win the election. His Centre-Left (PRD) opponent, Cardenas, was comfortably ahead when the computer system 'crashed'. When it started again, Salinas was miraculously ahead, and was declared the winner. Gabriela watched it in shock, seated in front of her TV.

The 2006 presidential election is widely believed to have involved fraud by the party of the Neo-Liberal candidate, Calderón, from the right-wing PAN. As yet there is no official proof of this, for the same reason as in 1988. In 2008, there was another Mexican election which involved widespread fraud. This time, it was an internal election for party president of PRD. The two main candidates accused each other of fraud, of which accusations at least some have been proven. Our aim in examining these two elections, 2006 and 2008, is to illustrate paradoxes of non-linear causality by comparing their contradictory outcomes.

In the internal election of 2008, the struggle was so bitter it factionalized the party. No authority was accepted to resolve disputed votes or arrive at an agreed count. As soon as one level of the hierarchy took action to resolve the issue, at the behest of one faction, the other faction appealed that outcome. However, in the 2006 Presidential election, electoral authorities resisted calls for a recount. They succeeded in imposing the result, which was probably a fraud, and certainly needed further scrutiny.

In 2006, the system still limped on at the national level. Its credibility was damaged but its power remained intact. Paradoxically, corruption was greater in this case, since the electoral system itself was compromised. Yet these greater levels of corruption protected the corrupt. Within the PRD party, the fraud affected lower levels of the system, but had worse consequences. The party was paralysed, unable to resolve its problems. It was more damaged electorally than the party and president who probably won by fraud in 2006. The more absolute the corruption, the more it can protect itself.

There was another paradox noted by commentators. The PRD was the party which suffered most from Calderón's alleged electoral fraud in 2006. Having suffered in this way, people asked how they could do the same thing themselves just two years later. One commentator wrote:

> It is no accident that the conflicts around the internal electoral processes of the PRD involve old members of the PRI. It is impossible to forget that Arturo Núñez was the top electoral official in the State Department when 'the system broke down' in July 1988, an outcome very similar to what the PRD electoral committee adopted when the flow of results surprised their expectations. (Loaeza 2008:27)

Loaeza traces the political process. When the PRI lost power, she suggests, many who used to hang on to it looked elsewhere. They joined other parties, its former opponents, PRD, PAN and others. They also brought with them an ethos:

> stuffing ballot boxes, stealing them, changing electoral rolls, kicking out political opponents... They did these things because that's how they are, and because they believe that that's politics. (*Ibid.*)

Loaeza had a metaphor for the process, drawn from a popular saying: a '*mala yerba...*', a noxious weed, which propagates with dangerous speed. The saying continues '... *nunca muere*', never dies. In Chapter 3, Zoë used the metaphor of a retrovirus. In Gresham's terms, this is 'bad money' which drives out good. It does so especially if the authority of the state insists that 'bad' is 'good', that a fraudulent candidate is the legitimate winner. Discursive corruption accompanies other forms of corruption and magnifies their effects.

One of our interviewees threw light on how these transactions can work out in everyday practice. Marcela works with the new reformist PRD government of Mexico City, dedicated, she said, to tackling corruption in the city administration, closely associated with the PRI regime that had monopolised government for over 70 years. For her, this made PRD officials especially sensitive to anything that seemed corrupt. She illustrated this with a story:

> This case concerns a more or less close relative (the sister of my father) who wanted me to 'recommend' her daughter's husband for a position in the government. Apparently this person was unemployed since he had been a member of the PRI and remained without office when the government changed. This person was *muy transa*, (slang for 'tricksy'), and I said to my aunt that I wouldn't be mad enough to recommend this man, since he would surely be a problem, given that the government now is PRD and they are trying to show that they can govern without corruption. My aunt was very annoyed.

This little story encapsulates the dynamics of resistance to reform in a process of transition from an embedded system of corruption. This man was cast adrift from the system he had been formed by all his life. He was no doubt desperate, like anyone made redundant after years of service. He cannot be blamed for his wish to be employed by the 'new firm'. Yet the dangerous mindset has also continued. The aunt seems unable to grasp the fact that those values are now unacceptable.

A *mala yerba* still grows in the mind of the person she recommended, seeking a new place to take root. There are many like him. In this case he did not succeed, but he would not stop trying because, as Loaeza says, 'that's how they are'. Many others like him have infiltrated other parties and other areas of government, PRD and PAN alike.

We will model this persistence of corruption by using two scientific concepts, *structural coupling* and *drift*, developed by Chilean biologists Humberto Maturana and Franciso Varela (1988). Structural coupling, they say, occurs when a history of interactions between an organism and its environment leads to mutual adaptation. Over time, they say, both systems drift, changing in step with each other, in a process of co-drifting co-evolution. Each is partly adapted to the other, responsive to its changes. Together they form something like a single organism.

This idea pushes us back in time, to see what structural coupling of behaviour and environment has led to current forms and problems. Mexican historian Claude Lomnitz (2000) argued that practices of corruption in Mexico's present reach back in a continuous chain to the colonial period, when Mexico was a colony ruled at a distance and on the cheap from Spain. There is a similar direct line for Brazilian corruption.

Agents of the crown were expected to provide results (wealth, order) with resources which were too few and came too late, following rules that had to be disobeyed in order to produce the required results. In this respect they were larrikin-like. They had to extort moneys in order to carry out their function, and then extort more, to make their profits. In this way they were more like managers in charge of a permanent downsizing operation.

Corruption colonial-style required processes of governance adapted to these circumstances, to take advantage of the contradictions and ambiguities of the formal system for personal gain (Lomnitz 2000:15). These contradictions and ambiguities were already there in the environment, symbiotically linking the agents of the crown with those who served them. This environment rewarded and required what is now called corruption.

All Latin American countries are products of European colonisation, which continues to be influential today. The *mala yerba* which currently damages political life in Mexico did not arise in 2006, with Calderón, or in 1988, with Salinas. The fraud and corruption which are damaging Lula's Brazil today was rampant in the Brazil of his modernising predecessors.

Neo-Liberal conditions and policies do not cure corruption, as its proponents claim. Nor are they solely responsible for the problems of corruption. But Neo-Liberalism does create conditions under which it flourishes, especially in the name of fighting it.

Corruption

'Corruption' is so often misused that it is tempting to try to do without the word. Neo-Liberalism takes it over and partitions it off into the strategic contradictions of its ideological complex, blaming all the wrong people, making 'good management' seem like all that the world needs. Yet at deeper levels it acquires more systemic, intractable contradictions. Corruption is the dark side of living systems. It attacks crucial sites of vulnerability, of every system and relationship. The effects of acts of greed and deceit multiply catastrophically, causing distress and death. Yet, changing the metaphor, this is also where dark matter resides, the non-luminous everyday world where the Larrikin Principle grows and thrives. The different kinds of darkness often look very different in day-to-day life and judgement. But sometimes, in difficult conditions, it can be hard to tell them apart.

Chapter 6

Power versus goodness @ the edge of chaos

In this chapter, we continue our examination of our key themes, Neo-Liberalism, Managerialism and the Larrikin Principle. This time we shift our strategy and focus in order to stay most of the time with Brazil, one of our three bodies. Each body is a site where the common patterns occur in unique configurations at different fractal levels, giving an ever richer and more complex understanding of the themes.

We will follow Fernanda on her journeys to the place and country of her birth. Belo Horizonte is a large city to the north of Rio de Janeiro. Brazil is an interesting site in many ways for 'first world' citizens to study the processes of globalization, and the policies and ideologies of Neo-Liberalism in extreme conditions: at the edge of chaos, where destruction and creativity become more evident.

Brazil today is full of contradictions. It is internationally famous as a symbol of exuberance, beauty and joy, yet it has high levels of violence and poverty. Brazilian sociologist Roberto DaMatta eloquently captured some of its contradictions:

> In Brazil we have carnivals and hierarchies, equalities and aristocracies, and cordial meetings full of smiles giving way the very next moment to the terrible antipathy and violence of 'Do you know who you're talking to?' We also have the samba, cachaca (sugar-cane brandy), the futbol (soccer), but this all occurs in the midst of what came to be called a 'relative democracy' and a 'Brazilian-style capitalism', a system in which only the workers run the risks and from which they draw no profits. (1991:2)

The disparity in income of the wealthiest 20% of the population and the poorest 20% makes it among the 10 most unequal nations in income distribution in the world (UNDP 2004). The International Labour Organisation estimates that 25,000 Brazilians work under conditions similar to slavery (Scharf 2008:72) Yet currently, it is one of a small group of countries collectively called 'Recently Emerging Economies' (REE), also known by the acronym BRIC (Brazil, Russia, India, China). In some longer lists of REEs, Mexico also squeezes in. In 2002 Brazil elected a Centre-Left President, Lula, from the Workers Party, but he disappointed many of his followers by maintaining many of the Neo-Liberal policies then in place.

For a long while Brazil has had an ambiguous relationship with the Neo-Liberal project in different forms and at different times. In the 1970s a

'Brazilian miracle' was explicitly based on the Neo-Liberal program; however, as Freidman objected, this did not include fiscal restraint. The prosperity it produced was impressive but short-lived. It created major social problems, including the massive destruction of forests, environmental degradation, and uncontrolled population movements to the cities.

In Belo Horizonte, Fernanda's home town, this led to the creation of a thick swathe of slums, illegal settlements and other sub-standard public housing around its edges, called in popular discourse *a periferia*, ('the Periphery'). The Periphery of Belo Horizonte is a fractal of Brazil in the world system. In spite of Brazil's recent success as an REE, it has not been able to clean up the scandalous conditions of the Periphery.

Brazil has a global impact in another way. In a Gaia map of the world climate, Brazil plays a crucial role. The Amazon basin, called 'the lungs of the world', is losing its forests at a dramatic rate. For instance, during the last five months of 2007, more than 3,000 square kilometres were lost (AEDT 2008). There used to be a saying, that if New York sneezes, the world catches a cold. It could equally be said that if Brazil's forests succumb to the cancer of uncontrolled development, the world should quit smoking, fast.

1. The problems of goodness

In 1992 Rio de Janeiro became important to the global world for another reason, when it hosted a landmark international conference, the United Nations Conference on Environment and Development, (UNCED). This conference became the focus of global concerns about the environmental crisis.

In 2002 Fernanda was drawn back to her home city when she became interested in a project in Belo Horizonte that we will name Verde (Portuguese for 'Green'). This was directly influenced by this conference, and by 'Agenda 21', the document that carried UNCED's blueprint for sustainable development. Fernanda was excited because this project was billed as a success story for environmentalism in Brazil, in her home town no less. Like most global citizens in various diasporas, larrikina Fernanda responded viscerally to the call of home.

It turned out that the story of Verde was much more complicated than Fernanda had expected, but in ways generating many reflections on causality on the edges of the edge of the global system. The project was established in 1995 with the aim of enhancing the quality of urban life in Belo, but the story soon became depressing. The project was officially wound up in 2003, after a frustrating struggle by its grassroots supporters failed (Duarte 2006).

Initially this seemed a sad illustration of the weakness of the good institutions designed to promote the wellbeing of the global system, when opposed to the big battalions that support the current system. It finished by offering an inspiring example of the non-linear effects managed by Brazilian versions of the Larrikin Principle. They achieved small and sustainable miracles in wastelands left by Neo-Liberal 'miracles'.

Globality

UNCED was a product of an apparatus constructed in the wake of World War 2, designed to correct the disastrous tendencies that had led to the horrors of that war. In effect this apparatus was a cybernetic loop designed to counteract the lethal global forces that had produced a great depression and two great wars, all larger and more global in scale than anything in human memory.

Robertson argues that these institutions and tendencies should have a different name than 'globalisation'. He calls them 'globality':

> A consciousness of the problems of the world as a single place [which] appears increasingly to permeate the affairs of all societies and multitudes of people across the world. (1993:132)

This term captures the positive aspects of *globo* we discussed in Chapter 4. It puts this complex, little understood set of processes on the agenda, as they deserve to be.

The institutions of globality sometimes seem so fragile as to be worse than nothing. For many 'anti-globalisation' activists, the forces behind Neo-Liberalism seem so powerful that they are the only effective reality, leaving 'globality' a mirage.

In this pessimistic view, its weak correctives will always be overridden by the overwhelming effects of those who run the Neo-Liberal system. For these pessimists they will be further negated by the chaos and incompetence of those on the end of the causal chain, on the margins of the global imperium.

Finally and most crushingly, 'globality' may exist only as an effect of business globalisation. For these doubters, it has been irreversibly coopted, so that it is now, and maybe always was, just rhetoric, discourse. Its only aim, in this view, is to sweeten the negative image of globalisation sufficiently to allow the reality to become even more monstrous.

Our task in this chapter is to use our analytical tools to tease apart some of the contradictions and confusions that make this despairing conclusion seem the only one possible. We will argue that globality is a set of emergent structures, dissipative systems forming at the edge of chaos. Globality is well adapted to survival there. Multinational corporate enterprises need globality if they are to survive. Sooner or later they find this out or perish.

Cybernetically the systems that make up globality are indispensable. Like Adam Smith's 'invisible hand', they do not rely on the goodness of anyone's heart. Cybernetic causality is non-linear. In Wiener's original metaphor of the *kybernautes*, the steerer, a small movement in the rudder transfers through many linkages into massive changes in the movement of the boat, constantly adjusting to turbulent seas. Globality may be as fragile as a butterfly, but as Lorenz said, butterflies can have large and unpredictable effects in far-from-equilibrium conditions.

We illustrate this with the complex histories of the International Monetary Fund (IMF) and the World Bank (WB). They began life as institutions of globality, to regulate gross imperfections in the world system, but in conditions where they were already dependent on the powerful, and governed by the logic of power. Their function was to provide stabilising feedback to correct the system, but they sometimes had opposite effects. Once captured by Neo-Liberalism they exacerbated the dangerous tendencies of the dominant system, with an ideological apparatus that concealed the fact of their failure.

William Easterly (2006) worked for the World Bank; he was an insider like Stiglitz and Williamson. He pondered on the record of global aid institutions, the bodies that made up the Washington Consensus, and came to a negative judgement. There has been USD 2.3 trillion in Western aid over the past 5 decades, he says. Their hearts are in the right place. But all this money has hardly touched the problems. After a lifetime in the Aid industry, 16 years in the World Bank, he came up with a depressingly simple analysis and solution, captured in the subtitle of his book: 'why the west's efforts to aid the rest have done so much ill and so little good'.

At the heart of the problem, he says, is the dominant role of Big Plans. He contrasts two approaches. 'Planners' have big plans, decided from afar, imposed from above. 'Searchers' work from the bottom up, trying to improve things in small, practical ways. Big Planners set 'beautiful goals', such as making poverty history, and then design agencies and resources to achieve it. They ask the question: what form of aid will achieve this aim? Searchers, he says 'ask the question the right way around: what can foreign aid do for poor people?' (2006:10).

He is moving in the right direction, pushed by many years of experience and reflection. So it is interesting to see that all his years with the World Bank have formed his mind so strongly that he still cannot turn the question really around, even when he says he wants to.

The preposition is still wrong: i.e. 'for' not 'with'. 'Poor people' are still on the receiving end of the Aid machine, just from a lower fractal level. The breakthrough question he wants but is still unable to ask is: 'What can poor people do with foreign aid?' Or more radically: 'What can poor people do, with foreign aid and without it?'

Easterly's ideas are still a radical challenge to the managerialist mindset which is the invisible heart of the international aid industry. The right plan, he says, is to have no plan. By that he means the managerialist idea of a plan, with clear objectives defined from above, specific steps and itemized deliverables.

We see his work, based on so many years on the front line of the Aid industry, as empirical support for propositions of chaos theory and the Larrikin Principle. Linear, top-down solutions to highly complex problems can produce worse outcomes. His 'searchers' are our larrikins. They cut through big rhetoric to get the job done. They listen and adjust their plans, working creatively with whatever is available, knowing and respecting the ideas and aspirations of local people.

Easterly says he does not doubt the idealism of Big Planners. As larrikins we are more sceptical. Yet he knows these people well. He was one of them. He is a reminder that the institutions of globality themselves are not intrinsically evil, run by cynical hypocrites, even though there is something seriously amiss with their basic assumptions.

The IMF and WB have become better than they were, he says. Stiglitz (2002) is less sure, especially about the IMF, though he sees improvements even there. For both, it is still possible for things to be done very differently. That is why we need to look at alternative causalities negotiated by various actors in the current global system. Globality may be threatened but it still exists. It still has non-linear effects that may improve the world in the ways that Easterly hopes for.

Globality at work

We begin our story with the UNCED conference held at Rio De Janeiro in 1992, which Fernanda attended from Australia. The brief for the conference was a definitive statement of globality, a global concern for the state of the globe. A brochure promoting the conference proclaimed:

> The world community now faces together greater risks to our common security through our impacts on the environment than from traditional military conflicts. We must now forge a new 'Earth ethic' which will inspire all peoples and nations to join in a new global partnership of North, South, East and West. (UNCED 1992:23)

Many problems of globality and its Neo-Liberal ideological shadow can be detected in this short, representative statement. The text constructs a global 'we', the world community united in 'our' common concern for no less than the Earth herself. The ethic is based in what is claimed to be a pragmatic, material problem; no less than the security of the earth.

But this 'we' is also responsible for 'our impacts on the environment'. 'We' includes Neo-Liberalism's devotees, people destroying the planet in the interests of profit, and consumers whose habits feed that profit. The problem is included in the solution, without any sense of where one ends and the other begins, or any strategy or route for going from one to the other. This is not productive fuzziness. This is thought in which all category differences have collapsed.

Links between analysis, exhortation and action are not explicit. We 'must' forge this new ethic, which 'will' inspire 'all people and nations'. Otherwise, nothing will happen. The good will is not in doubt, but the power and resources to make it happen are less obviously available. This has the form of an Ideological Complex again. This one blurs the difference between huge aims and meagre resources.

Fernanda returned to Australia after the conference, but some of the momentum continued in Brazil. The Verde program was a product of the enthusiasm generated at UNCED. In 1993, Belo Municipality successfully applied to the United Nations Development and Life programmes for funding.

There were conditions attached to the grant. 'Conditionality' has been a point of contention with Neo-Liberal funding, but this does not mean that there should never be any conditions. Paradoxically, that is what Neo-Liberals should think, according to the ideology they profess, though the Washington Consensus shows the opposite in practice. Conversely, critics of Neo-Liberalism should not see interventions and conditions as automatically wrong. It depends on what the conditions are, how they are arrived at (imposed from above, or negotiated from below) and how they are deployed.

In this case, one condition for the funding was that it should include a Non-Governmental Organisation (NGO) as a partner. In 1995, an agreement was signed between two departments of the Council, Environment, and Resources, with LEAF, our invented name for an NGO with experience in environmental activism, alternative technologies and community development. The program was located in four sectors of Belo, roughly North, South, East and West. In this and other ways it was a fractal, a mirror image of the global organization described by UNCED.

Mirror images do not easily fit into a linear model of power, and in spite of providing some of the funding, the global body had limited power. In 2002 Fernanda interviewed some key players in the Verde story. We begin our version of the story with the most senior bureaucrat Fernanda was able to interview. Miguel, a senior Council official, said:

> I have a vision of Agenda 21... They say 'Let's implement Agenda 21'. Well, I believe Agenda 21 is unimplementable. It'll never be implemented. It's much more like a code of conduct for action and planning than a program. So the impact of Agenda 21 on Belo can be explained in terms of a reflex of actions within public action, rather than only a program of work.

In the original Portuguese, as in our translation, this sounds good, but not entirely clear. It has the classic form of the Neo-Liberal/Managerialist

Ideological Complex, operating in a corner of Brazil. This senior bureaucrat distinguishes his 'vision' (the positive 'theory Y') from implementation. This will be managed through the instrumental rationality (theory X) that translates ideals into effective actions. But in this characteristic formulation, Miguel is using the contradiction to explain why nothing can be done. His managerialist can-do expertise identifies do-not-bother-with ('unimplementable') problems,

In this dualistic mode he was positive about the Verde program: '[It] for us is a very special program, firstly due to the fact that it has noble objectives'. This is the praise of 'beautiful ideas' Easterly saw as the mark of the Planner. But as Fernanda discovered in her research, the program was almost dead as he was speaking. The Council terminated the contract with the NGO, and cut all direct funding. In place of the financial and logistical support of the Council, a punitive funding scheme was imposed on the communities that still made up Verde.

Yet, in this interview, Miguel took a weirdly positive view of the situation and this history. On the stormy end of the relationship with the NGO he said: 'It was not a problem. The deadline ended. I think they became a bit complacent'.

In fact, as Fernanda found out, the NGO had had all its funding removed. They worked for no pay for many months till they had to withdraw. Miguel referred to one problem from his point of view:

> The public authority (Poder publico) was left out in the question of the process... It didn't have much influence.

The strategy he adopted restored control to the Council, without resources. The Council abolished the partnership, and made sections of Verde bid for specific funds for specific projects from a much smaller budget. This would guarantee the optimisation of resources, quality, monitoring, everything.

This is a caricature of classic Managerialism, including his unexamined faith that more competition with less or zero resources would do the trick. Yet as we have seen, something like this is standard practice for managers in the developed world who are facing 'downsizing'. Their restructuring/rationalisation requires fewer workers to do more with fewer resources than before, without any explanation of how the miracle is to be achieved. In Australia, this practice is so common that it contributed to the local name for Neo-Liberalism, 'Economic Rationalism' (Pusey 1992).

Miguel is clearly evading a reality he must know well. In cybernetic terms, he relies on not noticing what he does not want to know. Presumably he passes his denial further up the line, to top executives who do not want to know either, and do not want to take responsibility for not knowing.

He creates a discursive triumph to substitute for administrative failure. An enterprise without funding, as any competent manager or larrikin knows, will not optimise 'resources, quality, everything'. Optimising nothing is still nothing. We see here three ingredients of the dominant Neo-Liberal discourse as it appears in local practice. The rhetoric of environmental concern is yoked to a managerial discourse of 'implementation', and then negated by actual management practices. The outcome is a maximum of discourse and a minimum of action: exactly what the Ideological Complex is designed to deliver.

Instrumental reason in a chaotic world

Miguel's thinking illustrates what happens to mainstream management thinking as it circulates in wider and more turbulent times and spaces of global processes. As a bureaucrat he was trained in Brazilian forms of what the great 19th century German sociologist Max Weber described as *instrumental rationality*.

For Weber, this form of rationality was at the heart of the triumph of capitalism as a political and economic system. It was a key plank in the modernist paradigm that underpins capitalism as ideology and practice. Born in 17th Century Protestantism in Britain and America, according to Weber, the spirit of capitalism then evolved into the 'economic rationalism' central to Neo-liberalism. In this form it is vigorously exported to the 'developing' world in the name of 'globalization'.

According to Neo-Liberalism, the greatest barrier to global prosperity is the persistence of irrational practices in the developing world. These practices are presented as the major impediment to the take-up of the one true way of doing global business, and ensuring universal prosperity.

Weber saw bureaucratic rationality as the highest form of administration: 'Bureaucratic administration is, other things being equal, always, from a formal, technical point of view, the most rational type' (1947:337). He called it 'the most crucial phenomenon in the Western state'. It is the essential basis for the exercise of power, he claimed, thus explicitly linking order in one sense (orderly, rational arrangements) with order in another sense, the effective power to command.

In Brazil, bureaucracy and the notions of order and rationality play an ambiguous role. The national flag, inaugurated in 1889 at the establishment of the new nation, carries the motto *ordem e progresso*, order and progress. This phrase was lifted from the 19th century French positivist Auguste Comte, for whom order was the precondition of progress. But President da Fonseca, who chose this motto, was in power as the result of a military coup. He supported the interests of Brazilian elites rather than rational, efficient government. The gulf between motto and reality, flag and social life, is so great that it can make Brazilians wonder, longingly, what order would really be like, if it ever came to Brazil.

More than 100 years ago the Brazilian state declared that 'order' (instrumental rationalism) constituted national identity. Yet Brazilian everyday reality has never been characterised by order in any sense. A popular Brazil saying is *Manda quem pode, obedecer quem quer* (Those who can, give orders; those who want to, obey). This saying captures a deeply fissured society split between the will and irrationality of rulers, and the behaviours of the subordinate classes, between the arbitrary, intermittent power of the powerful, and the solidarity that is the essential but scarce basis for that power.

The result is a dysfunctionally over-elaborated bureaucracy, not imposed by agents of Neo-Liberalism but inherited from Brazil's unequal past. It is administered now by the elites whose 'self-interest' is the object of tender concern to Neo-Liberals, according to Williamson. In her fieldwork, Fernanda heard many critical comments from ordinary workers. 'Bums of lead', Carlos called the bureaucrats.

The ideal of 'progress' and the related slogan of 'modernisation' were also popular in Mexico from the 19th century, from the regime of Porfirio Diaz (1884–1911) which was only overthrown by the Mexican revolution of 1910–17. Porfirio's program of 'modernisation' attacked everything that was 'backward' in Mexican life, especially associated with Indigenous and rural people.

Neo-Liberal President Salinas in the 1990s promoted a revival of Porfirio's image and the idea of 'modernisation, through public education programmes and the national curriculum. As yet another paradox of Neo-Liberal rhetoric, 'progress' can refer to going backwards, not forwards, to find heroes of Neo-Liberalism in a rewritten past.

In Brazil, as in Mexico, Neo-Liberalism and its fractal shadow Managerialism interacted with the inherited dysfunctions to make matters worse. The funding crisis in Belo was a response to the indiscriminate cost-cutting that is a primary Neo-Liberal and managerialist strategy, the 'rationality' that defined 'Economic Rationalism' for its Australian critics.

Miguel's response to the funding crisis was to elaborate the bureaucratic apparatus, mostly with new managerialist mechanisms rather than older forms, but with a similar effect. The insult added to the injury, in this as in so many other cases, was to claim that this further impediment to function was actually a benefit that would compensate for reduced funding.

2. Pathologies of power

This environmental program was a micro-event that happened in Brazil, with a Brazilian flavour, but the problem is not unique to Brazil or the 21st century. In the 1940s, Merton, a follower of Weber, identified a systemic dysfunctionality in bureaucratic rationalism in its dominant Western (US) forms. He claimed that this form of rationality displaces organisational goals, something which in turn 'develops into rigidities and an inability to adjust readily' to new situations (1952:365). We distinguished between Brazilian and mainstream forms of bureaucratic rationality, only to find that the Brazilian case is only a more extreme version of a common problem. That makes it a convenient laboratory for exploring normal pathologies of power and management in exaggerated forms, in acute conditions.

Pathologies of trust

Verde, the community organisation, quickly came to seem ungovernable to the Council authorities, and this was seen, even by management, as related to issues of trust. One manager whom Fernanda interviewed called the environment 'toxic'. Francisco, a former director of Verde appointed by the Council, saw the program already in crisis when he began his tenure, only three years after it had begun. He identified two main causes. Money from the Council to pay salaries was already 6 months overdue, and Verde was out of the Council's control. The lack of control according to Francisco was partly due to the physical presence of LEAF, the NGO, and the distant, 'hands off' approach of the Council:

> Because the Council didn't have a physical presence in the community centres, I created control plans to monitor LEAF's activities. And this further complicated the matter, because things were very loose, very open. They didn't like the control.

However, unlike Miguel, Francisco was able to learn from the experience. Later in the interview he claimed that, in hindsight, he would have done things differently.

> I think in the area of the environment, if you don't engage in informal activism, you cannot advance. [There had been a need to] remove a bit [of] this structure of "government". Too many papers, too many authorizations that have to go through hundreds of people, only to release resources...

Francisco here has discovered for himself the main elements of the Larrikin Principle, as the outlines of a solution to his Brazilian problem.

The conflicts over control and issues of attitude raised by Francisco have been discussed in a recent debate about the nexus between trust and control. Bijlsma-Frankema and Costa (2005) reviewed recent sociological literature on trust-control theory, and came to a conclusion that surprised them, about the relationship between trust and control. They found as much empirical support for the idea that formal controls and trust are complementary (formal controls can enhance trust) as for the opposite idea, i.e. that they are inversely related (formal controls chase out trust).

A stark contradiction like this is not welcome in mainstream linear forms of sociology, especially if the opposing studies are equally sound, as seemed to be the case here. The contradiction is a warning that there is something else going on here, perhaps in the kind of control, or perhaps in the conditions where the two interact.

We will use the experience of Verde to throw further light on the issue. The story we have told seems at first glance to be empirical confirmation of the inverse relation between trust and control. The greater the control – the less the trust. But there is a further twist. Once control has chased out trust, the lack of trust then jeopardises control. The solution in this case, such as it is, shows that the same paradoxical principle can work towards improving the situation. When Francisco and other representatives of the Council modified the controls to make them more fuzzy, they were more effective. They made themselves more visible in the community and more worthy of trust.

Crisp control systems, as Zadeh's principle predicts, produce the opposite of control: irrelevance, or meaninglessness. Francisco's initial managerialist response (like Miguel's) to the 'crisis' in this far-from-equilibrium situation was exactly wrong. Instead of making the controls more explicit, formal and rigorous, in the managerialist way, they should have adopted fuzzy control systems, and tried to build trust.

Revealingly, in Fernanda's interviews, managers talked of 'control', whereas other employees talked more often of *confiança* and *credibilidade*: trust felt by workers, the credibility that inspired it. This difference of emphasis between management and workforce is a sign of the primary orientation of the two worlds in conflict.

In sociological literature, Portuguese/Spanish *confiança/confianza* is usually translated into English as 'trust'. Clearly these are equivalent terms from English and the Romance languages for a single fuzzy concept, and that is how we will mostly treat them. Yet the two words have different histories and etymologies, and their complex semantics carry traces of these differences.

Confiança ultimately derives from the Latin *fido*, usually translated 'I believe' (some person, or statement). Behind *fido* is Indo-European root *bheidh-*, to see or know. Related to this root is Greek *peitho*, I persuade. So in the Latin tradition, a form of knowledge is bound up with a relationship to persons who claim to know. Knowledge or belief underpins the relationship, rather than the other way round.

'Trust' comes from an equally ambiguous origin, Middle English *trewe*. This became modern English 'true'. Modern 'truth' applies more to statements than relationships, though love in popular songs is still 'true'. Middle English 'trewe' had the same two meanings, in a different balance. Behind *trewe* is an Indo-European root which produced English 'tree' (especially oaks and pines) and Latin *durus*, 'hard', 'lasting'. Just as oaks and pines are strong and lasting, what is 'true' can be 'trusted'. With *true*, belief follows a relationship of trust built up over time, while with *confiança*, trust follows belief.

Both lines of language weave the two qualities together, so they have become more or less equivalent. But the history of the terms makes qualities of knowledge more primary for romance languages, and qualities of relationship more basic for English. This is ironic, given that the Anglo world that has shaped the forms of language and thought of Managerialism now prides itself on its possession of 'objective truth', not influenced by subjective qualities, including trust. After the lessons learnt from the excesses of unfettered capitalism and the hard-sell in all aspects of corporate life, it has been forced to recognise the old virtues of trust that need time to build up.

Power and control

'Control' is the same word across all three languages. It ultimately derives from Latin *rota* (a wheel), which in Middle French became *rolle*, a roll of parchment. From this came the idea of a register of names, against which (Latin *contra*) actual humans were listed and 'controlled'. The idea of control, closely tied to modernity in many minds, is literally mediaeval in its origins.

'Control' began its life as a dualistic, linear model of power exercised through writing. It is intrinsically connected to bureaucratic power, power exercised through language, translating plans conceived in that modality into actions in the material world. Power is a fuzzy term, related to force as well as to social power. 'Control', in contrast, is more dualistic and linear.

In revisiting theories of the relation of 'trust' and 'control', we will put the two terms into the context of a more basic set of terms for social relations, power and solidarity. Here the French theorist Michel Foucault has contributed an influential legacy which connects both with chaos theory and the empirical problems of the Council.

Foucault is widely seen as one of the most important social theorists of the late 20th century. His ideas have already been applied extensively to the study of business and organisations (see, for instance, Clegg, Courpasson and Phillips 2006). There is much we find useful in his work. Yet as with every thinker, it would be limiting and unlarrikinlike to accept him as an unchallengeable authority speaking universal truths. When Foucault's ideas, formed in a European centre of thought, are encouraged to circulate in remote and different spaces, like Belo's *periferia*, strange and interesting things happen to them.

The idea of power as something linear acting from above is still the dominant idea of power in business, as for instance for bureaucrats like Miguel. Foucault rejected it:

> Power must be understood in the first instance as the multiplicity of force relations immanent in the sphere in which they operate, and which constitute their own organisation. (Foucault 1978:92)

Power in this sense is fuzzy (both 'power' and 'not-power', to some degree) and takes many forms, acting in many directions. It is different from the linear concept of 'control': '[Power] is never completely controlled from one point of view' (Foucault 1980:60). Foucault also proposed an intimate relation between knowledge, power and 'truth':

> 'Truth' is linked in a circular relation with systems of power which produce and sustain it, and to effects of power which it induces and which extends it. A regime of truth. (1980:133)

'Regimes of truth' certainly existed in Belo. Miguel for instance insisted on his right to define what has happened because of the power he believes he has, and he regards his knowledge as guaranteed by his position. What he says is the truth because he is the one who says it. This delusion is dangerously likely to be shared by his superiors, who will not seek out other versions, nor listen to them.

But Foucault's theory of power-knowledge is surprisingly close to what bureaucrats already think. From our larrikin perspective this is not a universal theory of knowledge and power. It is more like the common tendency of bad bosses to believe their own bull-shit.

The concepts of 'trust' and *confiança* add a further dimension to this complex, the quality of the social relationship, which 19th century French sociologist Emile Durkheim called 'solidarity' (1952). Durkheim diagnosed a breakdown in order and relationships as the characteristic pathology of capitalist forms of society, which he called *anomy* (lack of regulation or order).

Trust-control theory is presented as a relationship between two terms. Foucault also talks of a relationship between two terms, different but overlapping. We will put the two models together, along with Durkheim, into a three-body model, composed of three elements, power, knowledge and social relations, in a dynamic, ongoing relationship.

In Foucault's theory of power/knowledge, these two terms are totally enmeshed, so much so that they lose all separate identity or mode of action. Power becomes an effect of knowledge, and knowledge of power. But in a three-body model, the closure is never final and complete. Solidarity becomes the invisible but indispensable support of the power/knowledge complex, and also a casualty of it.

The three-body model supports our larrikin doubts about Foucault's idea of power/knowledge. This could not be a universally applicable theory of power and its relationship to knowledge, as many theorists have taken it to be. On the contrary, it is best seen as a pathological and unsustainable extreme, as Durkheim's category of *anomie* also is.

Viewed in these terms, this merging or collapse of crucial categories that we have seen in the case of Miguel and the Council is a common symptom, not a universal law. It shows what happens in extreme though perhaps common management situations. It shows the unpredictable, non-linear operations of power, knowledge and trust in far-from-equilibrium conditions. Like the discursive corruption we looked at in the previous chapter, this is a pathology.

It is this pathological form of power that operates in the situation of Verde in chaotic Brazil. As in the Brazilian proverb we quoted earlier, power ('poder' – both power and capacity) is ineffectual because orders in a chain of command do not reach those outside the circle of command. Putting the situations examined by control-trust theory in a framework of far-from-equilibrium theory, we can say that, in Brazil, power and control are dissociated.

But Foucault's theory of power does not correspond to the situation in Belo in one respect. He described one limit case, where the system of power is seamless and all-pervasive, occupying all networks, monopolising all forms of knowledge. The case of the community group, Verde, near the other end of the spectrum, seems to show the opposite qualities. Power, solidarity and knowledge are not locked into a single system. Power is not embedded in relations of solidarity or backed up by forms of knowledge, and it is separated from control.

Foucault described an ideal (or nightmare) instance of total power acting in what Prigogine would call a 'dissipative system', an open dynamic system which is highly stable in spite of the turbulence it has to deal with. Brazil largely lacks this form of power: hence its problems and the unpredictability of the outcomes. Yet these two conditions (and many more) are alternative possibilities within the same far-from-equilibrium conditions. They are opposites, but can flip over into the other in a moment.

This analysis has some implications regarding the operations of globalisation itself. If far-from-equilibrium conditions are distributed unevenly throughout the globe, this will profoundly affect the way Neo-Liberal structures of control play out between centres and peripheries. Even if the powerful became increasingly dominant, as many predict, this would still not guarantee stability of power and control. On the contrary, more power which destroys solidarity and truth will push the world towards chaos and ungovernability, globally, nationally and locally. That already seems to be happening.

3. Thriving in chaos

Most of us are brought up to believe in binaries, operating in a finite, linear world. From this point of view, power and goodness are opposites, and only the naive would think otherwise. So we expect the powerful (capitalists, corporations, big nations, politicians) to be bad, and we expect the good to be weak.

The opposite of this view is also prevalent in the modern world, especially in the discourses of Neo-Liberalism. This claims that the powerful are irresistibly powerful *and* also have a monopoly of goodness, to the only goodness which counts for them. They reframe it, colonise it, appropriate it, and make a profit out of it. At the same time they diminish and denigrate the goodness left over, which they cannot control or profit from.

There is a third alternative, which we believe is important to insist on. It is the inverse of Foucault's pessimistic knowledge/power complex. It envisages a reciprocal complex of power and goodness, where power enables goodness, and goodness sustains power. This again can only exist in far-from-equilibrium conditions. It is a dissipative structure. It is health. It also happens, and when it does it is worth nurturing.

In this section we will explore this other line of possibilities, while grounding our hope and optimism in empirical data. Many cynics confuse their cynicism with what they call 'being realistic'. That 'realism' ignores other aspects of reality at the edge of chaos where small miracles may happen, and do.

Three body systems and effective partnerships

One productive condition imposed on Verde and the Council was the insistence on three-way partnerships: the Council plus two other partners, a community organisation, and an NGO. This takes us to the notion of three-body systems proposed by Henri Poincaré. Three-body systems are inherently unpredictable. This makes them inconvenient as tools for linear planning, since after a while they will drift away from what was specified and predicted. But the same quality gives them planning advantages in conditions of uncertainty, at the edge of chaos. Like Prigogine's 'dissipative systems', they balance their unpredictability with a kind of stability.

We can see the merits of this arrangement in the trajectory of Verde. In the conditions of the grant, there had to be an NGO: hence LEAF's foundational role, as third body. This condition was a complication that the linear thinkers in the Council did not want. Initially, the Council itself involved two entities, two departments, Environment, and Resources. That would have created another three-body system. But Council squeezed out Resources, not because Resources were not involved (they were) but because they felt two departments would complicate matters.

They were left with a three-body system they were required to have, but Council management re-constructed this system as a pair of binaries. They contracted the function of running Verde to LEAF, the NGO, but remained distant from the community. LEAF became part of two binary structures: as agent of the Council, and administrator of the communities. Sandra, a manager from the Council, used a metaphor to describe the structure that emerged:

> It was a marriage in which LEAF believed that it was the Mother of the project, the Council the father, and the community the children. The mother was very kind to the community, but she was a bit possessive... The image of the Council for the community was one of an entity that doesn't deliver, doesn't pay salaries, delays everything... This was the image.

This 'image' was close to the truth, as Sandra later acknowledged. She sees it as dysfunctional, but does not reflect on why. In practice, this model for marriage is an alienated patriarchal family, its three terms (father, mother, child) split into two dyads, husband-wife and mother-child. In this image the dysfunctional family mirrors a dysfunctional organisation, at a lower fractal level.

In the event, the Council did not undergo marriage counselling. They simplified the structure by removing the inconvenient 'mother', even though they lacked the skills to carry out her function. In place of a partnership, they introduced a system of outsourcing, creating a multiple set of dyadic relationships entirely under their control, rather than a partnership approach.

Yet, in spite of the dead hand of linear systems, Verde did not die when funding stopped. Left to themselves, the remaining Council staff, all dedicated people, continued to be active on behalf of the ideals of the programme. In the words of Miguel, who was one of them, 'something interesting happened'. There was a 'flame that enabled Verde to keep going'. This was the local communities, whose ongoing relations constituted social capital and who still made use of almost the only physical capital left, the buildings and the gardens created by earlier projects. They also connected with LEAF as a virtual presence:

> We more or less follow the methodology used by LEAF. This was one of their positive points.

Knowing better what the communities wanted, not just imposing Council priorities as had happened before, Verde co-ordinators used their networks to find other potential funding partners, with great success. One was an international Foundation, drawing on multinational funding sources:

> I remember, when someone from the Foundation rang us to say the project had been approved, the resources, we had a carnival here [Laughs]. We had a big party!

Instead of the single oppressive relationship with the Council, with its linear model, its desire for control, and its excessive bureaucracy, Verde developed the fluid partnership approach that was integral to the initial scheme, and they did it with great success.

Even the final response, having a big party, carried a profound social meaning. The Brazilian love of parties does not necessarily show the superficiality of Brazilians, as it may appear to linear, protestant minds. On the contrary, parties help to build networks, and strengthen affective links that are essential to getting things done contrary to the worst efforts of a rigid bureaucracy.

Corporate Social Responsibility and the stakeholder model

One key to the success of the new Verde was the fact that it could step outside the oppressive, hierarchical, linear structures set up by the Council, and avail itself of a greater range of options in a more complex network model. This strategy touches on a number of themes in management and organisation studies. One is the notion of 'corporate social responsibility' (CSR) which lies behind the input of the above Foundation. The other is the notion of a networked stakeholder model, in which this and other inputs function.

Both these ideas have been critiqued in Critical Management Studies, for good reason. Both can be seen as typical examples of the managerial/Neo-Liberal ideological complex, combining a rhetoric of goodness to justify capitalists and capitalism, in order to leave business free to practise business as usual.

Indian/Australian management writer Subharata Bobby Banerjee has written strongly on this theme. Banerjee's irreverent book title *The good, the bad and the ugly*, makes use of a popular film. He explains how the terms apply to his theme. The 'good' refers to noble sounding corporate ideas and discourses:

> The bad is what these ideas and discourses conceal: how the imperatives of profit accumulation and shareholder value maximization do not always create win-win situations but often result in dispossession, and how the current political economy results in an economic capture of the social that marginalizes millions of people in the world. And the ugly is about how relentless corporate and government public relations campaigns create an illusory perception of good when describing the bad (2007:2).

Banerjee gives many examples to back up these claims. We also note his careful qualifications: 'not always' and 'often'. Yet Verde and the Foundation which helped fund it seem to operate in a space opened up outside the 'often'. Verde did not feel dispossessed. On the contrary, they held a party when they got the grant.

We broadly agree with Banerjee about what often happens in CSR. We only want to show the conditions under which the opposite can also happen. One key here is a cybernetic model. Cybernetics allows us to see two apparently opposite outcomes of any policy as both being possible, instead of supposing that a dominant tendency will remove all other options.

In this case, a crucial difference, between the kind of CSR Banerjee is talking about and this case, emerges when we consider issues of agency. CSR programmes driven by corporations are explicitly created to fulfil corporate purposes. As cybernetic loops, they may have a range of effects. On the one hand, they can change the organisation itself. More commonly, as Banerjee says, they allow corporations to continue as before, legitimated by CSR claims, to be as irresponsible as before, another contradiction of the Ideological Complex.

But in this case, the same CSR program is also an input into another system, the community organisation Verde. This system has far less economic power; but this does not make it less important for those who belong to it. Within this system, when the leaders of the community organisation exert agency, CSR resources may be the difference between viability and collapse. This may be true irrespective of the motives of the corporation, which may well be as self-serving as Banerjee says. A key problem will be constraints that are too tight, as they are in Neo-Liberal policies at the level of the nation, and often are at the level of the organisation.

In his critique Banerjee refers to 'shareholders', the sole drivers of capitalist organisations, whose sole motive, in one aspect of Neo-Liberalism, is profit. As Milton Friedman, the economic theorist of Neo-Liberalism, put it, with typical brutal clarity, 'The social responsibility of business is to increase its profits' (1970). Recently this linear shareholder model has been challenged by the more complex stakeholder model.

Banerjee gives a critical account of this model. He distinguishes between its normative use, to describe how corporations ought to behave, and its more common instrumental use, to carry out the primary business function of corporations better (2007:27). This captures the characteristic contradictions of the managerialist ideological complex. 'Stakeholder thinking' is an addition to the Theory-Y repertoire of softer models of relationships within business, and between business and society. Like other forms of Theory-Y it can be co-opted into an oppressive, schizogenic form of Managerialism. It can also be an element of the long-promised, long-delayed complexity revolution.

The term originated in the field of strategic management (Freeman 1984) as a way of planning strategies in complex environments: instrumental in that respect, but not necessarily an instrument to be used only by corporations. As with all instruments of management, larrikins are light-fingered with the Master's tools.

Stakeholder analysis brought a more complex form of planning into the limited, linear mindset of Managerialism. Other forms of stakeholder analysis have developed which are even more complex. Managerialism can invoke these as rhetoric, but they are too non-linear to be easily incorporated into strategic practice. Grassroots groups from below are not so linear or limited. Without the benefit of management theory, the activists of Verde mobilised all their networks. They replaced the corporation and different levels of government with themselves at the centre, in the kind of network model Rowley argued for (1997). They invented for themselves what Utting called 'multi-stakeholder' models (2002).

Creativity like this from below is a common manifestation of the Larrikin Principle. In Brazil, we saw it with Freire's theories, which empowered excluded, illiterate peasants to understand the nature and causes of their oppression, and change them. Still in Brazil, US anthropologist James Holton coined the term 'insurgent citizenship' (2007) to describe movements from below by the marginalised and dispossessed, who claim rights and responsibilities they have been excluded from for 500 years of rule by corrupt, incompetent elites.

In Mexico, the Zapatistas are the most visible form and focus of a similar kind of 'insurgent citizenship', but highly complex stakeholder strategies are used by many other community organisations. As one instance, the Fair Trade movement that is now a worldwide phenomenon grew out of the activities and initiatives of a Mexican community organisation, *Uciri*, over many years, building alliances with activist groups from many nations, themselves having many active networks in business and society.

In Australia, Corporate Social Responsibility was government policy even under the Neo-Liberal regime of John Howard. A policy document on the theme commissioned by the Howard government proclaimed:

> Corporate responsibility is emerging as an issue of critical importance for Australia's business community (Commonwealth of Australia 2006:3).

Banerjee would be sceptical of this, with justification, given its Neo-Liberal source. Yet in the realm of ideology, nothing can be taken for granted. The 'good' of CSR may still leak out of rhetoric into practice, in spite of the best (or worst) intentions of members of the business community. These intentions are diverse, belonging to an unmanageably complex set of active citizens with a stake in what they do.

Stakeholder theory emanating from above may be designed to fulfil only the purposes of corporations and managerialist managers. As a strategy, however, it describes a world too complex and non-linear to be limited and controlled in this way. Whatever the managerialist intentions, others can use it too. Such complex, multiple patterns of alliances and purposes are what the model is designed to show. They can and do happen in reality, too.

Thriving in chaos

The Verde project as viewed from the top seemed perfect but impossible to Miguel. But closer to the community level, where people had richer social networks, and less linear categories, the project looked far more successful, with different reasons for this success. Each of these people had multiple roles, multiple functions, but that did not present as a problem in itself, just a relationship to be negotiated.

Where the constant change in personnel was unsettling to Miguel as a manager, it was not a problem to Maria, who came from the community:

> It's alright to have new people in projects like this...I make a point of always bringing new people into projects.

Far from being disconcerted by the new range of possible partners, Carlos was eager to pursue many more. Miguel was nostalgic for the good old days when there was only one department of the Council to relate to:

> It's difficult to give continuity to a project without articulation with other departments, with the Regional department (which plays a coordinating role), resources, education, and health, with sanitation.

Milciades, a community leader, in contrast, was aware of the difficulties of participative methods, but saw many compensating qualities in the creativity of multi-body systems:

> Sometimes the process to get things done takes time, especially when you involve a greater number of people. But the importance is the quality. Things turn up that leave us surprised. Through the participative methodology there emerge ideas and incredible proposals that we would never expect.

All these lower-level members of the community were conscious of how much they had learnt, how much they had built up their cultural capital. For instance, Maria could speak the elaborated, abstract language of the bureaucrats at times:

> It would be good if the government embraced the work of urban agriculture... We're talking about conquering and using space, alternative technologies. Food security, the problem of waste. And it would be good if we could have the support of all departments, because urban agriculture involves other things, food security, waste disposal etc.

She is not parroting the words of better-educated officials. She uses the language of the 'tools of the Master' to frame her critique of the gap between the official rhetoric and the things that need to be done. She appropriates it; she is not being manipulated through it. From her position of non-power she dismantles the power/knowledge complex to assert her own power, enhanced by her own knowledge.

She has concrete aims, and grounded strategies:

> We have to get our will heard. Because everything we say is beautiful, but we have to have our feet on the ground and recognise it's a difficult and long process.

Maria and the other community activists have learnt huge amounts, and their organisation at their level has become a learning organisation in the way advocated by Garratt (2002) and others. They are good examples of Holton's 'insurgent citizens'. Unfortunately, the Council, with its linear communication flows and horizontal structures, was unable to learn anything from this episode. Their obsession with power was a barrier to knowledge.

To go back to the beginning of this causal chain, the pronouncements of the international conference in Rio ten years before, what is surprising is that Maria's response is very close to what Agenda 21 hoped would happen, but had no mechanisms for enforcing, and no right to expect. If the international body acted directly on Belo Council, it would only have been the upper levels of the bureaucracy, where as we have seen, people repeated the words while subverting the intent. Clearly the success story does not show linear causality, though it should also be emphasised that Agenda 21 and the funding it made possible did play a significant catalytic role.

We do not want to overstate the success the new Verde was able to achieve, compared to a well-funded and supported operation. The larrikin/insurgent citizens of Verde managed to create something remarkable out of almost nothing, but that something was not sustainable over a longer term. Their contribution was negated and finally reversed by the massive indifference of the dominant official system in Belo, ostensibly devoted to the same aim.

The power of goodness

The lessons about globality and the power or weakness of goodness are as mixed as chaos theory should lead us to expect. The ideals of globality will not effortlessly and inevitably prevail, in spite of all the material obstacles, against the overwhelming force of dominant systems, locally and globally. But nor will management methods and ideologies and their linear methods of control inevitably succeed.

The mechanisms and institutions of globality are products and channels of an 'invisible hand' which has more effects than seems possible to linear analysis. They still need and deserve much greater resources, to help create a more just society and a more sustainable planet. They also need a different mindset, closer to the Larrikin Principle, in which initiative and creativity from below are valued, where non-linearity is accepted as a fact and opportunity, and not just as a problem, and the larrikin sense of justice is the essential basis for trust. Without that, the world system and all its parts will come apart, and globalisation in all its forms will grind to a halt. No one really wants that.

Chapter 7

Soft Capital and the Informal Polity

Thus far, we have mainly used the Larrikin principle to organise our argument about organisations on a local and global scale. However, it would not be in keeping with the Larrikin Principle for us to insist on this term, as a new fashion or dogma, to replace all other terms in common use. The Larrikin Principle should not lose its dynamic flexibility, as a tool everyone can use to think with, from whatever background.

In this chapter, we explore relations between the Larrikin Principle and two other terms in general use, which cover some of the same ground. 'Social capital', popularised by French sociologist Pierre Bourdieu (1986), US sociologist Robert Putnam (2000) and others, has made a huge impact on social theory and management studies. The term in its different versions brings onto the agenda some key larrikin features, especially the rich, complex medium in which the Larrikin Principle operates, providing problems, resources and solutions.

The Larrikin Principle also assigns high value to informality, informal rules in informal systems, communication via informal channels. This set of practices poses a problem for linear theories aligned to dominant official systems, but they are so important in practice that they are to some degree acknowledged, although minimized as far as possible in Managerialism. However, from a chaos theory perspective, these are the rules and systems that are produced by and for non-linear conditions.

In this chapter we bring different aspects of informality under a single term, 'the informal polity' (see Coronado 2008), to emphasize the way different kinds of informal practice can hang together as a coherent system, underpinning and constituting a counter-group within a dominant system. Similarly, 'social capital' brings together a set of features that constitute a group, giving it coherence and effectiveness, as the cash nexus can hold members of an organisation together.

We provide a simplified map of some of the relationships we see between the three terms (Social Capital, the Informal Polity and the Larrikin Principle). Some of these details will emerge in the chapter to come. Here we comment on a few main points.

Firstly, we represent each of the three terms as a node at the centre of its own circle, with the circles overlapping to create fuzzy boundaries between them. Any one circle thus makes sense of the connections that make it up, and each is an incomplete picture of the networks of meaning that intersect within it. That means, in practice, that we need not be overly concerned about the limits of any one term, since it will connect with all the others (and more) through these networks.

Secondly, we represent a fissure in the fields of Social Capital and the Informal Polity to include the Other they are bound up with. Theories of Social Capital only make sense in terms of the operations of Capital in economic systems. Informal systems likewise are interconnected with the formal systems they supplement or oppose.

Thirdly, we diagram aspects of the three terms which are outside one or both the others. For instance we see challenges to authority and concern for social justice as qualities that mark the Larrikin Principle.

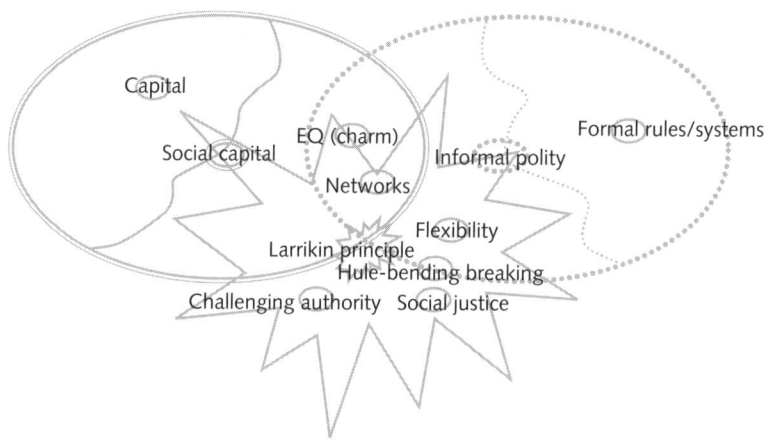

In this chapter, as in the others, we are interested in the implications of these concepts for business and organisation studies. Management writers recognise the existence of informal systems, and sometimes even their value, to some degree. Within Managerialism, that recognition is usually tugged back by the gravitational force of the dominant linear mindset, with its preference for unity and control from the top, and its commitment to the status quo.

Yet informal, soft systems are not an optional extra. Cybernetically, they are structural. Managers need to understand this complex, non-linear terrain, and need larrikin help to save them from the consequences of their linear practices. So do leaders of global capitalism, and every fractal level in between.

1. Larrikin correctives to dysfunctional systems

We will begin by using the Larrikin Principle as our organising node to look at soft systems and the informal polity. Again we begin our exposition in Brazil, reflecting on a field trip which Fernanda made for the project in 2006. Our concern is not to emphasise larrikin-tendencies in themselves, but to point out how larrikin qualities connect with qualities of social capital and the informal polity.

Brokers for broken systems

Fernanda made the following entry in her field work diary:

> It's after 5:00pm and I leave the office of two *despachantes*, following a very satisfying interview. As I walk down the ramp leading out into the street, I hear one of them say in a loud voice, as though he wants to be overheard: "*Muito simpática esta Fernanda. Ela sabe como usar charme p'ra conseguir as coisas*" ("Very charming this Fernanda. She knows how to use charm to get what she wants"). Sensing jocularity in his tone of voice, I stop and turn around to see both men staring at me with cheeky smiles across their faces. I smile too, because the man is right: I 'got what I wanted' – a very rich interview. Although the *despachantes* seemed a bit tense at the beginning of the interview – perhaps a bit suspicious of my motives – they warmed up to me as the conversations developed, and eventually were quite relaxed. Yes, I remember very consciously trying to be as charming as possible during the interview, to get the information I needed, knowing as I do as a native Brazilian, how important charm is in this society. (Diary, 13.1.2006)

This apparently simple, everyday text is very rich from our point of view, a well-placed periscope into Brazilian society and the complexities of informal, 'soft' systems.

These men do a job which has no close parallels in Australia or Mexico. *Despachantes* (from *despachar* to finish or resolve a problem) are people who work as "trouble-shooters" in Brazilian society. They are a Brazilian version of the Larrikin Principle, shaped by the specific forms of dysfunctionality in Brazil. Mexico has an analogous role, commonly called *coyotes*. As the term implies, *coyotes* are dangerous tricksters, not to be trusted.

Despachantes are paid by the hour for tasks such as queuing up for people in public service institutions and sorting out bureaucratic problems related to processes such as driving licences, electoral documents and identity cards. They are one of many cybernetic loops whose role is to correct the huge inefficiencies of Brazilian bureaucracy. Carlo, a *despachante*, gives this portrait of his role:

> You must be patient; you must listen to people's stories first. Things will happen at the right time, and you must have patience to listen. So, this is

the other side, you must be patient. In this business, it's pointless trying to make appointments; there's no such thing as appointments. When you're troubleshooting for people, things can happen by lunchtime, or later in the afternoon, but they can also stagnate for a whole day. So, the thing is to have patience... You have a commitment with your client. But they don't want to know about it; they just want things done: "I've paid you, and I have the right [to demand]". So, if the connection [with the client] breaks down, things get difficult. And it's here that you have to come in with *simpatia*, to persuade the client to wait till the next day. [Giggle]

To do this, they need a range of personal attributes. Patience is one of them. Standing and waiting in long, slow-moving queues is not as enjoyable as a samba. When they reach the head of the queue, they need another set of interpersonal skills. They need to establish rapport with the person in the office as well as manage the frustrations of their clients. They also need to project *charme* (charm) and *simpatia*, a quality ...

> where an individual is perceived as likeable, attractive, fun to be with, and easy going [and] behaves with dignity and respect towards others (Triandis et al. 1984:1363).

They also need to understand the bureaucratic requirements well enough to know where there can be exceptions, or faster, smoother routes through bureaucratic tangles. Even illegal ones, such as bribes, though Carlo does not mention these. Like larrikins, they need to be familiar with both sides of the office window. Only with that flexibility will they be able to do their necessary job.

With all this, it is still a job. In Brazil it has a semi-formal status, with its own union, yet it only exists outside the formal system to correct the limitations of that system, to help it limp along. In Mexico the same function is mostly done within the informal sector, by time-rich members of social networks.

The informal sector, in countries like Brazil and Mexico, has to be very large. Mexican economist Victor Soria argued that its growth in 1990s Mexico was a direct response to Neo-Liberal policies (1996). It is yet another paradox of Neo-Liberalism that this uncontrolled, unwanted growth is a consequence of Adam Smith's 'invisible hand', without the rationality that Neo-Liberalism expects from invisible hands.

Soft research for soft systems

Fernanda's ethnographic text was so rich because she used the softest of research methods to study these soft systems. We will turn our focus to the terms of the exchange itself, to show some of the issues that can be seen at work here, and the lessons it can tell us about soft capital and informal systems in practice.

Fernanda used two forms of research here. She interviewed these two men, in relatively formal exchanges, which yielded useful information. Her diary however could go beyond the interviews because of its informal status, as a private record of the total interaction as she experienced it.

As a participant and observer she could capture a wide set of messages. These came through non-verbal channels, in ways that were not controlled in advance. She saw them, including their facial expressions and other body language. They saw her, including her dress code and associated codes. All participants also took account of sound, especially tone of voice. These informal modes of communication are basic to human interaction, yet they are mostly excluded by formal research instruments.

Because this is an informal record, her comments can be subjective. She is a participant, not an impartial external observer. That is what gives a richness to the data that would otherwise have been filtered out. The men's reactions are responses to her, and they make sense to her as the social being that she is: 'a native Brazilian' as she puts it, 'a Brazilian woman' (of a recognisable age and class) she could have added.

These qualities are part of their common 'social capital', shared by researcher and researched. They are the background everyone involved draws on to create and interpret important social meanings. Social capital is both her means of research and its object. Without having social capital to spend she would not have co-created this event, or understood the social capital it contained.

The features she selects from their behaviours are signs that make sense to someone who is Brazilian and female. She reports their motives as she interprets them as though they were facts: both the loud voice, and the wish to be overheard. Yet that is something that happens in exchanges where motives are meant to be read. Fernanda is probably making the interpretation she is meant to make. That is how communication works in these conditions.

Surrounding the words is one informal mode, the tone of voice. Surrounding that is another level of the informal system, the shared culture. It does not need to be stated. It works best if it is not. When informal systems can be left implicit, activated by known triggers but not stated, they work more powerfully.

It was important that the moment captured in this report was an informal part of the exchange, deliberately outside the official, formal interaction. In the interviews, these men stayed mostly within the bounds of politeness. They told Fernanda what they chose to tell her in that context. Yet the attitudes they express or imply through this exchange are also what they think, and act on to some degree. In a formal, recorded conversation, they presented themselves as non-sexist. In the unrecorded exchange with Fernanda, they revealed the nature and scale of the sexism they feel, and no doubt often act on.

'Social capital' as a term makes connections with economic exchanges. In these terms we can say that Fernanda as a researcher wanted a commodity, information, knowledge, insight, which remained outside the cash economy. She did not pay them for their expert knowledge or their precious time. Nor did they ask her to. She in turn was not paid for the information, or not directly. She is employed by her university to carry out research, amongst other duties. But that obligation is vague, with no precise cash value attached. In spite of its managerialist ethos, the Australian university system still relies on informal systems and an informal economy.

Yet there is a transaction, which both parties are aware of and satisfied by. Fernanda was conscious when she planned the interview that she needed to offer these men something they valued for their efforts. She calls it 'charm'. So do they, so this is a well-managed exchange.

We guess that their last comment, the deliberately loud comment between them, worked to insert Fernanda into their own relationship, and its macho terms. The compliment to her implied their pleasure in her feminine wiles. The loudness could have been interpreted by Fernanda as a threat. However, on this occasion she smiled.

Smiles, like so much in informal networks of communication, can be highly fuzzy. She reports the meaning she gives it, personal triumph and complicity with the men. Equally important, it completes an unstated contract governing this exchange. Once she no longer needed their co-operation, she offered a bonus smile. Among other things, she signalled to them that her charm was not a mere stratagem, to be switched off as soon as she got what she wanted, even though as everyone knew it was indeed just that.

In this non-linear system of informal communication, what they give is worth more if it is really valued. Both parties are winners, and they win more the more the other also wins. This is not how economic capital works, but it is one feature of social capital.

The power of charm

The *despachante* used two words for what we translated as 'charm'. In his second sentence he talks of *charme,* like English 'charm'. Both come from Latin *carmen*, song, especially used in magic. It suggests something dangerous, irrational, irresistible. In this sentence it is a noun, a thing, a possession, a weapon Fernanda uses skilfully, or so the *despachante* claims.

The other word, *simpática*, is very common throughout Latin America, in both Portuguese and Spanish. It comes from Greek *sympatheia*, literally 'feeling with someone'. As noted earlier, *simpatia* implies treating others with respect and dignity. English 'sympathy' comes from the same root, but has a typically cool, British flavour, being sorry for someone.

There are differences between *charme* and *simpatia*, even though the *despachante* applies both to the same person, Fernanda, within two sentences. *Charme* is especially ambivalent, sometimes a positive quality, at other times seen as manipulation, to be resisted. English 'charm' has similar ambivalence. It does not sit well with the skill set of Anglo-American businessmen (we use the gendered noun deliberately). However, it has a positive sense in the self-help genre which people in business read avidly. For instance US writers Tracy and Arden published *The power of charm: how to win anyone over in any situation*. That is a big promise. They describe it as 'the ability some people have to create extraordinary rapport that makes others in their presence feel exceptional' (2006:4).

We do not despise this genre. On the contrary, we believe that Adam Smith's 'invisible hand' guides the books as they walk off the shelves. Many people know they need them. Dale Carnegie's best-seller of 1936, published as the USA was emerging from the Great Depression, was the founding father of the genre in modern times, selling over 5 million copies and still in print. The genre functions as a cybernetic loop, an informal curriculum to correct the rationality offered as the only route to power in management texts.

There is both an edge and a naivety in Brazilian uses of 'charm'. The web site Brazilista.com woos potential visitors like this:

> The average Brazilian knows the power of the smile and a smooth word and touch, which they use to their advantage, and can quite possibly leave the unsuspecting charmed out of their pants. So there, you've been served with fair warning!! (Brazilista.com, 2004)

Charm here manipulates, softly, deceptively. Its power may seem a threat, or maybe a promise, though larrikin readers might prefer more control over their pants.

Mariana talks differently about *simpatia* in everyday Brazilian life:

> This is something I've learned… You go to the bank, or to the movies, and sometimes people get into negative moods; they don't feel like working, or whatever. Look, what you must do is to act with politeness; be as poliiiite as you can [loudly and emphatically]. You may be feeling like shit, but you will smile; you'll say "Please"… otherwise people lose their good will, and you can just give up! That is, if you're not *simpática*, my dear, you'll most certainly hear a "No!"

This is *simpatia*, not *charme*, and Mariana shows the difference. She smiles although she feels like shit, to get something but also because she does empathise with people who serve her. What she wants is not a special favour, just whatever it is she is queuing for. Those she wants to persuade are resentful and likely to withhold this good, not because they do not have it or ought not to give it but because they feel so oppressed and powerless that this is the only power they have.

DaMatta describes the Brazilian reality that qualities of 'charm' in both senses play out against. Brazil is best understood as a 'dual social system', he says, which

> tends to feed itself on the dialectics of a Draconian impersonal law and a system of personal relationships that perversely enables people to get around laws and decrees (1984:10).

He uses 'impersonalism' for the official harsh, frustrating system which produces the resentful non-compliance of the oppressed and under-paid staff who are unwilling to serve people like Mariana, along with the tricks that she and others have to play to get something done. He calls 'personalism' the other part of the system, the informal systems, legal and illegal, functional and dysfunctional, that flourish in exotic abundance.

Like larrikinism, this set of systems allows people to get round laws and decrees, but the scale of the problems to be overcome lies beyond Australian larrikin experience. In Brazil there is anger and resentment on both sides of any counter or in any relationship between the powerful and the non-powerful. *Charme* and *simpatia* are not superficial qualities, but hard-won strategies to survive.

In Brazilian society, *simpatia* implies equality between people, a relationship of affinity. In this way it has similarities with larrikin *mateship*, though it is different in including personal attractiveness, like *charme*. Like *mateship*, albeit less so, it suggests a longer-term relationship. Jorgue is a customer relations officer:

> If a person is polite with you – face-to-face or even on the telephone – you feel that the person cares for you ... So, you end up creating a friendship with the customer who says "I found you very *simpático*." And then they come back, sometimes even bringing you something – sweets, or... or whatever. They don't need to do anything extraordinary; to tell you fairytales. You end up acquiescing. For example, sometimes I notice by the tone of their voice that the customer doesn't have the full amount of money to pay for the service. But I say, "That's alright. No worries." You just let go.

This customer is not a close friend, but the relationship is real. Gifts do change hands, but they are small, under the bribery-radar. If they were too big or too extravagant they would not work. He recoils from 'fairy tales', the equivalent of the larrikin 'bullshit'. In this gift economy, they act as signs of respect, which have to be felt as sincere, rather than simple inducements. In this environment they have non-linear effects, producing more than they are worth, or provoking a counterreaction if they feel excessive.

What is especially surprising in this exchange is the way Jorgue gets the message that a favour is needed. The customer does not plead. Jorgue 'notices' the problem, through tone of voice or other small signs. This person he likes, and who likes him, is in real trouble. *Simpatia* is a two-way process for him, not just something the customer has to show or pretend. He uses a classic Australian larrikin phrase to complete the process: 'no worries'.

The semiotics of *charme* are different. The basic relationship is unequal. The person with *charme* generally has less power, and needs to compensate with *charme*. The signs of *charme* tend to concern appearance, faces, figures, bodies, clothes, while the meanings they convey are about the immediate relationship, not a longer term one.

Yet both *charme* and *simpatia* are used in conditions where distinctions blur and get lost. In these far-from-equilibrium conditions, sex, love and friendship morph into each other, and appearance and reality can be hard to tell apart. In this environment *charme* becomes especially ambiguous, good or bad. It depends on whether you lose your pants or not, and whether you want to.

Tricks and dodges

To deal with far-from-equilibrium situations, Brazilians have developed a set of strategies called the *jeitinho*, which, we have suggested, have similar qualities to larrikin rule bending. The word is a diminutive of *jeito*, a way (of doing things): hence a short cut, cutting through bureaucratic tangles. The diminutive gives a familiar, affectionate flavour to the word. Mexican Spanish does a similar thing with *-ito*. So a *favor*, asking for special treatment , becomes a *favorcito*. Australian English does the same thing, adding *-ie* to make rule-bending seem harmless. Australian workers who take a day off for personal reasons, but still claim 'sick leave', call it a 'sickie'.

DaMatta relates *jeitinho* to the dual structure of Brazilian society:

> It is alarming to see that daily legislation in Brazil is a regulation of the "can't do"…It is precisely because of this that we discovered and perfected the *jeitinho*, a style of social navigation that is read between the lines of these peremptory and authoritarian "can't do's". (1984:98)

Javier, a senior Brazilian public servant, saw the *jeitinho* in a similar light:

> Well, I think the *jeitinho* only happens when there's a personal connection. I'll only help you if I think you're nice [*for com sua cara*]. If I don't, you'll need to follow the law. "To friends, everything, to foes, the law" [old Brazilian adage].

For Javier the *jeitinho* needs a personal relationship, or the illusion of one, in a simple, reductive structure which reduces social complexity to two broad groups, friends and enemies.

This scheme is so simplistic that problems arise. Most people do not know each other, as they would in an imaginary village community. Sign systems are likely to produce irrational, contradictory outcomes. Javier sees the law, the formal system, as inherently hostile, a way of doing bad things to enemies, and by which enemies do bad things to you. Laws are not about maintaining justice. They are not a common resource, available to all. As social capital they are the creation and preserve of hostile forces. In Chico's version of the Larrikin Principle, justice and close personal networks are incompatible.

2. Social capital

Practices like the Brazilian *jeitinho* can be understood in relation to any one of the three terms of this chapter, social capital, informal polity and the Larrikin Principle. As they combine, they provide a richer, more nuanced picture of these complex sets of practices. Yet there is also value in taking each of them as a central node that relates to the others, between them and with other aspects of social and organisational life. In this section we take 'social capital' as our central node.

Versions of social capital

The relationships involved in *jeitinhos* have been described by Brazilian sociologist Livia Barbosa (1992:34) as 'diffuse reciprocity', since people can receive benefits from *jeitinhos* they may not have given. Her idea of diffuse reciprocity sees *jeitinhos* as inherently complex. It also connects with the idea of social capital (see Duarte 2006) developed in different versions by French sociologist Pierre Bourdieu (1986) and others, including Coleman (1988), Putnam (1995, 2000) and Cohen and Prusak (2001). 'Social Capital' is one of those ideas which replicates like a retro-virus, mutating as it goes. There were 4,380,000 Google hits for 'Social Capital Theory' alone when we looked in late 2008.

We will not offer a comprehensive account of the ways the term is used. We limit ourselves to some framing comments. It is important to recognise that this is a metaphor, something which gives it a degree of complexity that many disregard. The meaning of a metaphor is dependent on two terms, and the relationship between them. This makes them an elementary three-body system, which increases the variety and instability of the term's meanings. In this case it puts a special focus on different meanings of the hinge idea, capital: how particular theorists understand it, how closely the metaphor is tied to those meanings, and what happens when as large and fuzzy a term as 'social' is added to the complex.

Another source of diversity in versions of the concept is the pervasive fact of contradiction, produced by different cybernetic systems or ideological complexes. 'Social capital' may be seen as something whose role is to correct the ills of mainstream capitalism, while others see it as a mirror or support of standard capitalism. The Neo-Liberal and managerialist ideological complexes include both versions among their functional contradictions.

For Centre-Left French sociologist Pierre Bourdieu, Social Capital reinforces effects of economic capital. It is

> the aggregate of the actual or potential resources which are linked to possession of a durable network of more or less institutionalised relationships (1989:249).

Bourdieu proposed an analogous form of capital, 'cultural'. The 3 forms of capital are transformed, via specific institutions, into what he calls symbolic capital:

> the power granted to those who have gained sufficient recognition to be in a position to impose recognition (1989).

All four forms of capital combine together as an accelerating feedback loop which multiplies the advantages of the already advantaged. Bourdieu as a social theorist was interested in the forces that create inequality, amongst which is wealth. Management is more interested in the forces that produce wealth, and hence contribute to capital. Bourdieu was a fierce critic of Neo-Liberalism (2003) whose version of social capital emphasises the force of one half of its ideological complex.

The currently most popular use of the term follows US sociologist Robert Putnam. He uses the term to criticise what he saw as a moral failure in the USA, the heartland of world capitalism. In cybernetic terms he sees 'capital', with 'social' attached, as a corrective feedback loop, to moderate the pernicious effects of capitalism generally, including Neo-Liberalism, on social cohesion in its US heartland.

Bourdieu and Putnam seem to contradict each other, but from a cybernetic perspective they are complementary. Their different ideas regarding social capital describe different aspects of a single, highly complex process, in which both kinds of social capital co-exist and interact, in sometimes unpredictable, chaotic ways.

Bourdieu and Putnam are both social theorists. We are also interested in how this kind of idea is taken up by authors more closely connected to the world of management. Again we see a polarity between different versions of social capital, reflecting different relationships to the Neo-Liberal ideological complex. Coleman (1988, 1990) reframed the idea of reciprocity as if it were an exchange carried out by the pure rationalists invented by rational choice economic theory:

> When I do a favour for you, this ordinarily occurs at a time when you have a need and involves no great cost to me. If I am rational and purely self-

interested, I see that the importance to you of this favour is sufficiently great that you will be ready to repay me with a favour in my time of need that will benefit me more than this favour costs me – unless of course you are also in need at that time (Coleman 1990:309).

Coleman applies his analogy strictly:

> If A does something for B and trusts B to reciprocate in the future, this establishes an expectation in A and an obligation on the part of B. This obligation can be conceived as a credit slip held by A for performance by B. If A holds a large number of these credit slips, for a number of persons with whom A has relations, then the analogy to financial capital is direct. (1988:103)

Anthropologists have explored the idea of reciprocity in complex forms in non-monetary cultures, following Mauss's classic study (1966). Coleman filters it through a model of rationality that no anthropologist would see as relevant to non-modern cultures, and which is hardly more relevant to societies today.

Rational Choice theory has been a basic premise for the brand of Economics underpinning Neo-Liberalism. This economics school has been criticised for its use of such fictions. Coleman offers no empirical support for his fictions, because there is none. Why did it seem so 'reasonable', to him and his readers, that he did not notice how far from reasonable it is? This is partly due to the ideological effect of the idea of 'rationality' within Economic Rationalism, though it is guaranteed only by the pun on rationality.

Against his fantasy we offer our own. Imagine a person who went round the office calculating the worth of favours done by him. He writes them down on slips of paper (with cash equivalent, estimated on some unstated basis). Then he tries to 'redeem' the slips with actions by others, again translated into a cash equivalent. Even in first-world organisations, someone who tried that would seem a hopeless sociopath.

Cohen and Prusak (2001) had a different idea of social capital, aligned to a different use. They start from their sense of how organisations ought to be, against the dominant managerialist model:

> Every manager knows that business runs better when people within an organisation know and trust one another – deals move faster and more smoothly, teams are more productive, people learn more quickly and perform with more creativity. Strong relationships, most managers will agree, are the grease of an organization. Business gets done without them, but not for long and not very well. (2001:86)

This has the form of Theory Y, with all the inherent ambiguity of such theories. On the one hand, social networks are part of a system that corrects Linear Managerialism. They produce creativity and efficiency, harmony and productivity. But Cohen and Prusak say that 'every manager' knows this. If this were so, there would be no need for larrikins and other mediators, no need for deep change. Managers already know this truth, and can be left to act on that knowledge. The revolution has already happened.

Cohen and Prusak are perhaps being strategic here. They propose real changes in managerial behaviour:

> Networks of social connection, trust, and commitment cannot be manufactured or engineered, only encouraged. (2001:23).

These qualities, they imply, exist and have their value independently of corporate spin. They will not work if they are faked.

The Larrikin Principle connects in significantly different ways with these various theories, so much so that it can be used diagnostically, as a way to understand some major differences. For instance, Bourdieu is strongly critical of Neo-Liberalism, but his theory of social capital has no clear place for the Larrikin Principle. Without a larrikin form of agency, he can show how oppressive and hostile to human values the dominant system is, but not where energies may be mobilised against it.

Putnam confronts the inherent failures of humanity of the dominant economic rationalism as a reason that it will have to change. By using the term 'capital', with its positive associations, he is able to mount an argument for change that Managerialism can easily understand, though the extent of the change is limited, being able to be accommodated within the ideological complex. Many of the qualities he and Cohen and Prusak include in social capital constitute the 'dark matter' the Larrikin Principle draws on.

Against these three versions, Coleman's ideas illustrate how far Managerialism is from grasping what matters in this body of ideas. It applies a linear, instrumental logic to the social domain, to include it completely within an accounting model. His hero, going round the office collecting slips, is poles apart from the larrikin. The implications of his theory for practice would be to turn all informal mechanisms into formal systems, as managerialists since Taylor have tried to do. We doubt that it would work.

Marx on the 'capital' in Social Capital

All these theories of Social Capital use analogies with capital in the economic sense. In this section we want to invert the analogy, and look at an account of economic capital, and the implications it has for something like 'social capital'.

Marx's work is a useful starting point for our reflections for many reasons. Marx drew extensively on classical economists like Smith and Ricardo, and surprisingly many of his basic ideas are still part of a consensus in all forms of economics. Especially useful for our purposes, his theory of capital included within it a theory of social capital, and an account of how the two are typically related under capitalism.

Equally usefully, he framed his ideas within an implicit semiotic framework. He treated money as signs or symbols. This enables us to talk of social factors like *charme, jeitinhos, favorcitos* and *mateship* in comparable terms, as complex symbols which play a role in social actions and relationships, working together with or against money capital.

He begins his explication of capital with a simple formula for exchange: $C \rightarrow M \rightarrow C$, where C stands for Commodity values and M for Money value. The pre-condition for the creation of capital, he says, is a market. It has to be a world market he said in the 1860s, long before globalisation became a buzzword. Capitalism began in the West, he says, in the 16th century, so it already had a long history when he was writing. In all this, he agrees with Adam Smith, claimed by Neo-Liberals today as their Founding Father.

In a first stage, the existence of money allows the circulation of commodities, underpinning the 'wealth of nations', in Smith's phrase. But this is not yet Capitalism for Marx. He then considers another kind of exchange, apparently much the same: $M \rightarrow C \rightarrow M$.

This includes the same elements, and can co-exist with the first: however, Marx argues that it is a fundamentally different process. In the first, money frees up the exchange of commodities, from which everyone benefits, as Smith said. From the second, traders aim to increase their money by making a profit: $M \rightarrow M_1$ as Marx calls the second M. That second M can buy more commodities, and produce more money indefinitely. It is now part of capital, and there is no limit to its productivity.

Marx's ideas on money anticipate Prigogine's account of the role of catalysts in the mechanisms of chaos theory. By as early as 1836, Berzelius had identified the remarkable property of catalysts, substances which speed up the rate of chemical reactions without themselves being affected. This is a non-linear cause and effect, discovered before the idea of non-linearity had been invented.

Prigogine took this idea further, connecting it with the role of enzymes in biological systems. Without enzymes, he says, metabolic processes as we know them would not happen, and without metabolic processes, life itself would not exist. He identifies three main forms:

> Autocatalysis (the presence of X accelerates its own synthesis), autoinhibition (the presence of X blocks a catalysis needed to synthesize it) and crosscatalysis (two products belonging to two different reaction chains activate each other's synthesis) provide the classical regulation mechanism guaranteeing the coherence of the metabolic function. (1984:153)

The way money breeds money ($M \rightarrow M_1$) is an example of an autocatalytic loop. It increases the rate of commodity exchanges while also increasing its own production. On its own, according to Prigogine, this process would produce runaway, unsustainable processes. Marx also noted unsustainable tendencies in the capitalist monetary system. In Prigogine's terms, it needs complementary autoinhibition loops, or it will burn out.

To make this kind of point, Marx drew on the classical Greek philosopher Aristotle, who distinguished 'economics', the art of gaining a livelihood, from 'chrematistics'. Aristotle invented this word, from *chremata*, money, the art of making money:

> For chrematistics, circulation is the source of wealth... Therefore wealth, as chrematistics strive to attain it, is unlimited'. (Quoted 1967:139)

Aristotle separated chrematistics sharply from economics. Marx agreed in principle but insisted that the two kinds of exchange are inseparable in practice. In terms of chaos theory we note that it is in chrematistics that processes accelerate to infinity. Its power laws produce a kind of chaos. In the composite field formed by 'economics' and 'chrematistics' under capitalism, chaos spreads from the money system and throughout the whole. The 'real economy' (Aristotle's 'economy') plays the role of autoinhibition in this composite system.

Crosscatalysis, as Prigogine called it, is also important in understanding the world financial system, and the crises to which it is prone. As is widely acknowledged, a non-financial, non-commodity force often referred to as 'confidence' has catalytic effects on both the other systems. In 2008, George W Bush, Neo-Liberal ('Free market') US President, faced with a domino effect in the financial sector, in effect agreed with Aristotle and Marx rather than Neo-Liberal economics. He made massive reserves of funds available to prop up failing banks. His main target was the panic that threatened to overwhelm Wall Street.

President Bush was not listening to soft-left/'liberal' bleeding hearts, but to people like John Mack, CEO of merchant bank Morgan Stanley. Mack wrote this despairing memo to his staff:

> From: John Mack
> To: All Employees
> I know all of you are watching our stock price today, and so am I. After the strong earnings and $179 billion in liquidity we announced yesterday – which virtually every equity analyst highlighted in their notes this morning – there is no rational basis for the movements in our stock or credit default spreads. What's happening out there? It's very clear to me – we're in the midst of a market controlled by fear and rumours, and short sellers are driving our stock down. (White 2008:26)

CEO Mack is railing against the irrationality in the supposedly rational market, although fortunes are made as well as lost by this kind of irrationality. But the fear is a kind of negative capital. It translates into huge flows of money, demonstrating both the irrationality of the system, its proneness to chaos, and also the role played by non-monetary flows in emptying out capital. Mack appeals, futilely, to the commodities in Aristotle's 'economy'. These ought to protect against 'chrematistic' forces, but do not.

In his memo he also asked employees to spread the good word to people they knew. In this way he was calling on other kinds of capital within his organisation. Symbolic capital, in Bourdieu's sense, had been emptied out with a rapidity Bourdieu did not discuss. Mack tried to counter outflows of emotional capital ('confidence') by appealing to the social capital, in Putnam's sense, that he hoped existed in Morgan Stanley. This included the positive feelings of 'all employees' towards him and the company, multiplied by their links to all their own networks.

To mobilise this capital he takes some elements of the Larrikin Principle. He addresses them personally, as 'John Mack'. He is one among equals, sharing their experiences as they share his. He states his aim in simple, direct terms, with no 'bullshit' He assumes that he has high levels of social capital with these employees, based on the relationship of trust that is an indispensable component of both good management and the Larrikin Principle.

Whether or not Mack had as much social capital as he claimed we do not know. Perhaps he was trying to cash in Bourdieu's symbolic capital instead. Either way, it was not enough to be effective. In the far-from-equilibrium world of chrematistics, emotions and feelings crosscatalyse to money, and money turns into nothing but those emotions, fear and distrust, which are the wrong ones, from Mack's point of view. But this is not a transaction a good banker can control.

In this account of 'capital' we are not talking metaphorically, about something analogous to economic capital, applied to a different system, as is the case with most versions of 'social capital'. 'Capital' in the sense we use the term here has a broader scope and takes more forms than purely financial capital. It has economic causes and effects, just as financial capital also has social causes and effects. The various forms are so interrelated that they constitute a single, highly complex non-linear system, requiring a non-linear theory to understand it, in its parts as well as in its whole.

Within this framework we can gather the non-financial forms under the term 'soft capital', to capture the sense that it is not easy to reduce them to hard, crisp terms and precise monetary equivalents. They are a force not to be denied, in markets and organisations alike. Or we could call them 'dark capital', to emphasise how they are largely denied by linear forms of economics and management, carried through informal systems that run alongside or drive official, formal systems.

Social capital and the cash nexus

Our expanded, differentiated sense of 'capital' includes many forms of capital connecting in many ways, in a single complex system. Yet the relation of money and financial forms of capital with all the others has problems which we need to touch on here. Countless transactions in formal and informal relationships of the social world are enabled by autocatalytic loops, like *charme*, which behave in many respects like money. Yet, in other respects, they behave differently.

Money has evolved into a public system in which values are to some degree supported by specific mechanisms, and this is essential if it is to accumulate into capital, on the scale that has happened under capitalism. It is not so clear that these other forms of transaction can accumulate into capital, on a large enough scale to become a comparably powerful force. Nor is it clear what happens as money, the dominant form of capital, interacts with these other forms.

Marx, among others, stresses the crucial role of money in capitalism. Capitalism expands the number of things that can be treated as commodities, and therefore bought and sold. Money is a universal agent of transformation, he says. Capitalism

> has torn asunder the motley feudal ties that bound man to his "natural superiors", and has left remaining no other nexus between man and man than naked self-interest, than callous "cash payment". (Marx and Engels 1968:38)

Brazilian Lorenzo in our corpus three-quarter agrees with Marx, but insists on one crucial difference:

> There are many things that you can only get with money – this is more than proved all over the world because money can get things moving. But there are people who'll try to buy with money, what they can get with a smile.

Lorenzo and Marx have a broadly similar analysis of the problem. In essence, chrematistics has spread too far. But, paradoxically, Lorenzo is the one still holding out, insisting on the continuing relevance of non-cash values. Marx is so caught up in his critique that he supposes that what he most hates has already come to pass.

Marx argued that the value of commodities was created by the human labour that created them. In this he followed once more in Adam Smith's footsteps. That made human labour and social relations of production part of the value of commodities. That was what capitalists bought when employing labour.

For the purposes of his argument, Marx put his main emphasis on human labour, 'congealed' into capital as he put it. That was to focus on his argument that profit, $M_1 - M$, arose from the surplus labour of workers, who produced more than they were paid for. But from his discussion it is clear that it is not just their labour that is exploited. The meaning of the commodity includes relations of production as well as labour. So social relations are also transformed into commodities, and hence into money capital.

In this sense we can say that, within it, financial capital contains social value, i.e. social capital. Without this element, in which workers form work teams and obey their supervisors, the job would not be done and there would be no profit. So Mack not only tried to draw on his employees' social capital in the crisis. It was the basis of business as usual, nurtured by Theory Y practices.

For Lorenzo, smiles function in a similar way to money. They incorporate relationships from the past. Smiles need to be natural, growing out of experiences of other positive relationships. They produce a relationship in the present, a bond of weak but productive goodwill. They are autocatalytic loops, which speed up the relationship, and, incidentally, produce more smiles.

Modern businesses in the Neo-Liberal heartlands of the USA and Australia are fully aware of the value of smiles. Everyone in sales is required to smile at customers till their jaws ache, and then keep smiling. They repeat social formulae like: 'Have you had a good day?' It is even better if they are attractive young women or men. All these aspects of 'charm' are required by bosses in Australia, as in the rest of the 'developed' world, and are included in the price of the commodity. But this is not usually itemised or rewarded, and frontline sales staff who have it in abundance are often not well paid. In this way what could have contributed to the employees' social and economic capital is appropriated, along with their labour, into the profits and thence the capital of the employer.

Another way of looking at what is happening here is to recognise that the work and product have to do with meanings. For this reason we can call this semiotic surplus value. This then builds up into semiotic capital, another buried part of general capital. Marx did not have a systematic position on semiotic capital, but he noted many instances of it in passing. For instance, he reported the findings of a commission of inquiry into bread in London in 1862, which found that three-quarters of bakers were 'under-sellers'. They sold bread below price because they adulterated it with soap, chalk, etc. The public paid the price for what 'bread' means, but got a cheaper substitute. Profit came from expropriated meanings, what would today be called fraud. These activities are forms of discursive corruption, as discussed in Chapter 5, and the generalised theory of discursive corruption we presented there is part of this extended theory of capital.

In this picture, the Larrikin Principle represents an autoinhibitive feedback loop that builds in informal social networks and shared realities. It attacks 'bullshit', all inflated claims, including the operations of Bourdieu's 'symbolic capital'. It dismantles the seductions of the Ideological Complex, especially the aspects which replace material benefits with feel-good words and images.

In this process the Larrikin Principle homes in on an important truth about Managerialism. The tricks and stratagems to coopt employees have structural causes and effects in contemporary capitalist organisations. They do not supplement the linear official system. They are part of all actual systems, attempting to translate them all into capital to be used, as far as possible, by and for the bosses. This is what the Larrikin Principle recognises, and resists.

3. The informal polity

One defining feature of the Larrikin Principle is the importance of informality. Larrikins are laid back, meeting the needs of a job rather than a pre-existing formal set of specifications. They communicate effectively through informal language, backed up by informal, non-verbal communication. Social capital, as we have seen, uses informal means and networks. 'Informality' binds our two categories together.

Yet there is value in specifically focusing on informality. In practice it refers to many areas. It raises issues of linearity and its limitations in a rich, comprehensive form. As a term it is so low-key that it is easy to minimise its importance, in every aspect of organisations. In this section we try to do it some justice.

Management texts on informal networks

It may seem that we do not need to argue for the value of informal systems with managers today. No management textbook we looked at is negative about informal systems or communication. We do not need to work hard to persuade anyone. We are mainstream, it seems.

Well, not quite. In a sample of 5 management textbooks, all include the concept of informality, as organisation or communication, or both. We give the percentage of each book that deals with informal organisations or practices:

Book	Informal org	Informal com	Total informal
Davidson	0.62%	0.5%	1.12%
Robbins	0.44%	0	0.14%
Hill	0.42%	0	0.42%
Waddell	0	0.39%	0.39%
Hitt	0	0.75%	0.75%

197

These figures show that the concept of informality exists in some form, but also that it is not prominent. Combining both categories, there is an average of 0.62%. Only one text book has more than 1%. That text is the only one to deal with both organisation and communication. Otherwise, these authors deal with only one or the other.

We found a much greater importance for informal networks among our interviewees:

Nation	Positive informal	Negative informal
Australia	29.6%	15.9%
Mexico	38.0%	72.0%
Brazil	27.0%	5.1%

These figures suggest that informal systems are important for people in all three nations,. But there are also some interesting differences between them. Our Brazilians were the least positive about informal systems, though not to a significant degree, and were the least negative about them. The Mexicans had the most positive responses, but were also by a large factor the most negative, with many cases of ambivalence. This shows a strong focus on informal practices, overlaid by acute ambivalence. In contrast, Australians were positive, to a degree, with a definite undercurrent of negativity.

One Australian, Mathew, a senior manager at a media company, gave his views on informal systems:

> I think they're crucial, they're really important and I think, yeah, working relationships really operate at lots of different levels. There's obviously the kind of official process level for which supposedly things like friendships and relationships and those types of things shouldn't really matter because everybody is professional and has a job to do and a process to follow and trying to achieve a common end but in fact there are so many other layers that meet that where most of the communication and most of the kind of deal making and negotiation and those types of things actually occur at the kind of informal network level I think. … it's just much more pleasurable and productive for everybody if you have that kind of work environment I think

Ray, another Australian, added his reflections:

> I think they're essential in the workplace and I guess in my own mind I make a difference here between informal practices and breaking rules, although one may very much go with the other. But certainly, informal networks are essential to help people survive, to help them see how organizations operate, to help them perhaps to get things done more efficiently, more effectively, if only from the point of view of being given the correct information at the right time. So yeah, I think they are essential for not only an individual's survival in an organization, but if you looked a little more deeply, perhaps for the

continuation of the organization. Those informal networks are sometimes necessary.

There is a sharp contrast between the treatment of informality in management textbooks and these reports from the coalface. Informal systems, we suggest, are easy to pay lip service to, in the managerial ideological complex, but too complex to include in a formal curriculum. In fact, they are treated as unteachable. There are no accompanying exercises, no examples reproduced in any of the textbooks. Either of our interviewees would be a better starting point for students to reflect on than anything (or more precisely, the nothing) in these books, and presumably in the courses they are part of.

Informal systems and the idea of social capital

Our own research asked questions about intersections between our three terms to different people across our three national sites. This aspect of organizations emerged with special richness and strength in our data. Our interviewees were asked to tell stories about how they or others solved problems arising from bureaucracy. Most of these stories (96.99%) included a solution, as we asked. The overwhelming majority of these solutions (85.16%) referred to an aspect of informal systems.

Even more interesting than this remarkably strong reference to informal systems as a source of solutions was the breakdown of the components. We sub-divided the informal polity into 3 categories: alternative systems, networks and rule bending. The following table provides information about examples of each in our data:

Node	References to informal polity	% references to informal polity
Alternative systems	272	43.04%
Networks	473	74.84%
Rule bending	405	64.08%

In 75% of our stories, networks were important. That agrees with Social Capital theories in the Putnam line. But 64% of the stories also used rule-bending in some way, something that does not figure in most social capital theory. Less common but still an important resource was the use of alternative systems. These are systems which exist outside the dominant system, yet are still rule-governed and legitimate for those using them to get things done.

We were interested to see how networks were created or sustained in the informal polity. The following table gives a breakdown:

Node	References	% ref/ networks	% ref/informal polity	% ref/ stories
using charm	90	19.0%	14.2%	11.8%
using status	65	13.7%	10.3%	8.5%
using gifts	57	12.0%	9.0%	7.5%
With stable networks	224	47.4%	35.4%	29.3%
With family networks	47	9.9%	7.4%	6.1%

The networks used to solve problems in our sample mostly drew on existing relationships, people already known in the workplace or beyond. Gifts were sometimes used, but networks are mostly mobilized, not created, to solve problems.

Rule bending is a crucial category in a larrikin perspective on the informal polity. We found three main forms of rule-bending in our data.

Strategy	Refs	% refs to rule bending	% refs to inf polity	% of refs all stories
Creativity	81	20.0%	12.8%	10.6%
Defiance	155	38.3%	24.5%	20.3%
Flexibility	214	52.8%	33.9%	28.0%

'Creativity' here refers to new ways to get around the problem. 'Defiance' refers to rejection, sometimes angry, of the rules in question. 'Flexibility' involves elements of both, but is characterised by adjustments between rules, problems and solutions.

A major distinction here is between direct opposition to rules and orders, defying the authority that tries to impose them, and two variants of a softer, less confrontational approach. In our data a significant proportion, 38%, involved direct defiance. A larger number involved just flexibility, in how far to apply a problematic rule, or how to get around it. A smaller but still significant proportion (20%) went beyond flexibility to involve a creative element, solving problems by means that had not been thought of, or forbidden by existing authorities or rules.

One important feature that emerges from these many examples is the sheer diversity of what we are calling informal systems. They are best understood not as a single system that operates as an alternative to the formal (official) system, but rather as a multiplicity of options arising spontaneously in practice. They are immediate solutions to immediate problems.

This is what makes them challenging to forms of management that give high priority to the consistency of the official system, and treat any departure from it as a threat to the integrity of that system. Yet many of the worst problems arise because of the rigidity of the boundaries between the official, formal process and the many practices that arise in order to allow tasks to be carried out.

In our sample, one quarter of respondents felt they had to defy the formal system, but three quarters found ways around the problems. This is not a picture of good management being white-anted by uncontrolled rule-breakers, but the contrary. It suggests that good management is often achieved precisely by respecting alternative solutions. It may also suggest that in many cases where rules had to be defied, it was the formal systems that were at fault, needing a larrikin approach.

The governance paradox

We finish this section by looking at strategies at higher fractal levels, where governments relate to the people they govern through formal and informal systems. We take our examples from Mexico, to illustrate the role of informal systems in a Neo-Liberal client state.

Neo-Liberal policies and pressures have interacted in paradoxical ways with the Mexican informal polity. Conditions on loans to the Mexican government since 1988, the fraudulent election of President Carlos Salinas, regularly included the standard conditions of the Washington Consensus: low tariffs, privatisation of public assets, reduced taxes, redirecting public expense towards modernisation, creating infrastructure that will attract big capital from international investors, away from social welfare for the majority of people. They also had to rein in the informal or underground economy.

There are some contradictions in this brief, which hinge on the role of the informal polity. In Chapter 4 we saw the contradictions in President Calderón's Neo-Liberal budget and in his defence of it. The same set of contradictions has played out in other Mexican government initiatives at other times. We call it the Governance Paradox.

The paradox is that governments need to use the informal systems which they also have to condemn and attack, in rhetoric and action. If they were to succeed in eliminating the informal polity, the economic and political system would collapse. The first duty and necessity of a government, the survival of the governed, would be jeopardised. The Neo-Liberal rhetoric, that its prescription is 'tough love', good for the country as well as for foreign capital, would be exposed.

On the smallest scale, one of our Mexican interviewees described why she sells goods from her home, although it is illegal. This is part of the 'underground economy' that the government is trying to stamp out:

> Josefa: Why are they going to fine me, if I am selling from my house because I do not have a job? It is not that what I am selling has been stolen...

Her victimless crime is created only by a system of regulations. This action allows her to survive without calling on the public purse (a good Neo-Liberal principle) while evading regulations and not paying tax (against Neo-Liberal principles). The costs of enforcing the rule in a systematic way would outweigh any revenues, with no clear public benefit. The benefit of not enforcing it, in this and countless other cases, is that Josefa and her family can get by without causing anyone any harm.

A different tactic of Mexican governments is to appropriate an informal system that emerged of its own volition from below. One example of this was a programme called *Mercados sobre ruedas* (markets on wheels). These itinerant markets sprang up in many Mexican cities, where poor people bought and sold goods, mainly in poor areas. Rather than close these down or move people on, the government declared it a government programme.

'Markets on wheels' was based on a thriving informal practice, which grew out of an earlier indigenous form of itinerary markets called *tianguis*. It transferred them from the rural areas where they originated to serve a similar purpose in cities filled by large numbers of those rural citizens who had migrated to them.

In this way, government policy legitimated what people were doing anyway, taking advantage of the creativity from below. The government dealt with the problem of reducing unregulated informality without having to provide something in its place, simply by declaring it legitimate. In this way they solved the problem that created the strategy in the first place, at almost no cost.

This is a beautiful example of Adam Smith's 'invisible hand'. The outcome was a highly complex, functional market. It was sponsored by a Neo-Liberal government pressured by the dictates of Neo-Liberalism to regulate it. It could not do this, because it could not otherwise cope with what typically follows from Neo-Liberal conditions. Instead, by doing almost nothing, it created (or more exactly took credit for) a thriving free market.

In a similar vein, an unofficial policy regularised land 'invaded' by large numbers of former peasants, moving from rural areas depopulated by policies that favoured agribusiness. These hard-working landless people moved into cities like Mexico City, where they took over unoccupied land, building houses and other amenities.

After a period of unsuccessful police harassment, the government of the day changed its tactics. They incorporated this practice into the system by claiming control over the purchasing of the land, then selling it to squatter families for a nominal price. In this way, the pressures of these landless people were met in practice, while preserving the veneer of legality and government control. The invaders acted in every other way as good citizens. They built roads and communal facilities which the government might otherwise have been asked to do.

The government probably would not have provided these things, so the informal polity did not save any actual money. However, it protected the invisible social capital which sustained its legitimacy, to some degree, as a government. Regularisation was already happening de facto before the government dispensations, through bribes to corrupt police officers and other officials. So in this way also, the government accidentally fulfilled Neo-Liberal conditions, (reducing corruption) by not following Neo-Liberal policies.

Soft capital and the informal polity

From the point of view of crisp logic, we should by now have chosen between our three terms, soft/social capital, informal polity and the Larrikin Principle. We see many advantages in not choosing. No society would be sustainable if it relied only on formal systems, consciously formulated, directed and enforced from above. This is as true of developed nations like Australia and the USA as it is of 'developing' nations like Mexico and Brazil. It is as true of teams and offices as it is of multinationals and nations. The existence and importance of informal systems is a universal law. It is a larrikin check on the arrogance of power at all levels, in Managerialism and Neo-Liberalism alike.

This law prevails because social life is a non-linear process that unfolds as the outcome of the acts of countless individual agents, following mobile and flexible larrikin patterns that balance the individual and the collective. The product, at every fractal level, aggregates to a stable set of resources. We liken it to capital, to capture its fluid transformations of so many diverse things into social forces, its vulnerability to exploitation by the powerful. Or we can use another term to encompass it, as we do in our next chapter: culture.

Chapter 8

Culture and organisations in a global world

In this chapter we use 'culture' as our final site in which to explore our core themes. Larrikin practices come from culture in one sense, Australian culture primarily. These cultural forms have a long history, whose twists and turns explain much that is otherwise surprising and contradictory in the larrikin. 'Culture' is an older and in some respects better way to capture what is called 'social/cultural' capital.

In business circles, top managers with big plans often see 'culture' as the site of resistance to their vision. 'Culture' is used to indicate that this resistance is deep-seated, irrational and backward-looking. The 'good culture' that will replace it is also irrational, but this is not emphasised, since this irrationality is harnessed for the corporate good. This still leaves a sense of unease around the term.

It also describes something found in nations from the 'developing' world, whose alien, incomprehensible ways are also seen as needing change. 'Culture' is convenient shorthand for all the ways in which they will resist the superior wisdom of the developed world. This wisdom is often seen not as 'culture' but as eternal, universal principles for how things should be done. 'Culture' is something that 'primitives' have in abundance. It makes them seem colourful, exotic and incompetent. 'Advanced' peoples no longer need it, except as a luxury.

In this chapter, we explore these takes on culture from business and management literature, including critical management studies. We examine some causes and consequences of these ideas in business. We also place this limited set of ideas of culture in a wider framework, to develop richer, more useable ideas of culture.

Ideas of 'culture' in the business world are typically filtered through the linear logic of Managerialism. In the process, much of the complexity drops out. We reframe ideas of culture through a chaos and complexity framework. This allows us to weave some beautiful, powerful ideas and analytical tools into the interdisciplinary resources of people in business. It is not that the idea of 'culture' is wrong. On the contrary, it is too good an idea to be lost or misused. Its complexities make it more valuable than ever in the global, trans-cultural world of today.

1. The many faces of culture

The key terms we deal with are complex, fuzzy and full of contradictions, and this is just as true of 'culture' as the others, in its own particular ways. We will look at how it functions in business discourse and business thinking, to bring out contradictions that need to be acknowledged there. We will also set up a wider framework, to see old problems and new possibilities with this term for thinking about business and organisations today.

Culture in organisations

We do not need to argue for the importance of 'culture' in business and management thinking. It has been mainstreamed in management for three decades. For instance, Burrell and Morgan (1979) developed a typology of forms of organisation which reflected different structural features, producing different kinds of organisation, different cultures, with different possibilities for change. Business gurus Peters and Waterman gave 'culture' a central role in good organisations:

> Without exception, the dominance and coherence of culture proved to be an essential quality of the excellent companies. Moreover, the stronger the culture and the more it was directed towards the market place, the less need was there for policy manuals, organisation charts, or detailed procedures and rules (1982:75).

In cross-cultural management, Dutch-US writer and management consultant Geert Hofstede popularised a concept of culture applied to organisations in different nations which has proved highly influential (Hofstede 1980, Hofstede and Hofstede 2005).

An interest in culture is well developed in critical management studies (e.g. Alvesson 2002). US management theorist Linda Smircich wrote:

> Complex organization as a cultural form has enabled us to provide universal education, to eliminate deadly diseases such as polio and smallpox, and to explore outer space. Complex organization as a cultural form also facilitates environmental destruction and the possibility of nuclear annihilation. A cultural framework for analysis encourages us to see that an important role for both those who study and manage organization is not to celebrate organization as a value, but to question the ends it serves. (1983:355)

Smircich insists that business culture is complex, and productive. At the same time she positions analysts of culture outside that culture, able to criticise it. That is necessary for a critical management approach, yet it is also a problem for analytic practice: how to find a position that is sufficiently outside the practice to see its limits, yet still inside it, to some degree, in order to understand it better.

Unlike 'informal systems' and 'corruption', 'culture' is common in management textbooks. All our textbooks cover it. Hitt et al. (2007) are typical in how they see 'culture':

> Ultimately, as manager you need to thoroughly understand culture because it can help accomplish your managerial responsibilities. So how does culture do this? Because culture is rooted in assumptions and values, once established it guides people's behaviours without overt or constant supervision. (2007:114)

There is nothing warm and fuzzy about this justification. Culture is like an implant. Once inserted into workers it programs them, so that they no longer need supervision.

But there is an obvious catch in this idea of culture. What if the wrong program is inserted? Hitt et al. recognise the problem:

> Culture is such a strong force in behaviour that the wrong culture can lead otherwise good people 'to do bad things' (2007:115).

The quotation marks around 'to do bad things' are symptomatic and revealing. Who are they quoting? They do not say. These are often called 'scare quotes', with good reason. This sentence opens up the scary prospect that everyone has ample scope to learn many forms of 'wrong culture', national and organisational. For all its attraction the term arouses ambivalence.

The term is used confidently in popular business journalism. For instance *Management Today*, magazine of the Australian Institute of Management (a professional body for Australian managers), carried a story on this theme in 2008. Its main article, by editor Jason Day, featured David Widdows, Australian CEO of MNC Heinz, reflecting on 'culture' as the key to his success with the company. The story describes how Heinz Australia was having a 'lean time' when Widdows took over in 2003. Widdows blamed the culture (2008:8):

> "Internally, it was a very negative 'that's not going to work' type culture," recalls Widdows. "The rewards for success were probably [outweighed] by the penalties of failure; a punitive culture, is probably the best way of putting it."

Widdows reports that he took two measures in the first two months. He sacked 25% of the salaried staff, and restructured the board. He then attacked the culture problem:

> A saying was agreed upon: 'A great place to work' and everyone was asked for input on what a great place to work meant for them.

Widdows 'wanted everyone, from himself down, to walk the talk. The outcomes have been very positive' says Day, and recognised at Head Office.

> "Over time, there's been a growing appreciation from the global CEO and others that, in Australia, culture is a real resource, and that culture *is* about results", says Widdows. (2008:9)

The role of 'culture' in this sense is not just a personal discovery by Widdows or Day. As we saw, Hofstede and Peters and Waterman had already beaten them to this discovery. Hitt et al. and other management textbooks had embedded it in the management curriculum, just waiting to be rediscovered by Widdows and his peers in the business world. Widdows is not an original theorist. Just as useful, he provides good data on how the word and its accompanying ideas are used.

Widdows reports a 'strong culture' in Peters and Waterman's sense, which also has some features of larrikin informal culture. Rules are not emphasised and bureaucracy is stripped down. But there is no sign of the egalitarianism, scepticism or resistance to bosses of the Larrikin Principle. On the contrary, this is a culture of focused conformity, driven by the boss.

Even in this superficial examination of 'culture', as it makes sense of aspects of organisational life, we can see the complexity of its meanings and uses. For Widdows and the culture of this magazine, it is both positive and negative, old yet new.

Uses of culture and the management paradox

Widdows uses 'culture' much as Hitt et al. and Peters and Waterman described it. From this we can see why the term is so useful for senior managers. It gives complex thinking more space in the managerial skill set. 'Culture' empowers Widdows to try a synthetic, non-linear approach, drawing many aspects of the situation into a unified judgement. In the common distinction in the literature, it allows him to present himself as a 'leader', not just a 'manager'.

As we noted in Chapter 2, Managerialism is ambivalent about leadership. We quoted the symptomatic puzzlement of Davidson et al.:

> Organisations need sound management, yet they look for great leaders. One could be forgiven for feeling bewilderment. (2003:565–5)

For these writers, management may be the same as leadership, or it may be different. Business puts leadership above management and pays the huge salaries that scandalised the post-crisis world. But even in 2003, Davidson et al. were feeling that there was something wrong with this. Clearly there is something disturbing, problematic, yet desirable about 'leadership' for Managerialism.

They have good reason for seeing tension here. In his classic work on forms of authority, Weber distinguished between rational/bureaucratic authority and what he called 'charismatic' authority (1947). Weber saw these as different in kind, appropriate to very different conditions. The bureaucratic leader has the qualities of a good manager. The charismatic leader is inspirational, intuitive, holding the organisation together, often only temporarily, by force of personality. They are such distinct qualities, valuable for such different circumstances, that they are not easily found or needed in the same person.

Yet, however different they may seem, there is pressure, in Davidson et al. and in Widdows' self description, to find a way to have both. The two sets of qualities reflect contradictions of the managerialist ideological complex. This would not be a problem if CEOs only had to pretend to be leaders, while remaining managers. This story suggests that Widdows must do more than pretend. Shadowing the contradictions of the managerial complex are similar contradictions confronting his practice. We call this the management governance paradox.

The terms of this paradox are as follows:

- Organisations need leadership *and* management, non-linearity *and* linearity.
- Leadership is incompatible with *and* an essential complement to management
- Organisations must repress and devalue the Larrikin Principle at every level *and* value it, even or especially at the top.

The term 'culture' plays a key role in allowing Widdows to negotiate this contradiction. He does not refer to quantitative methods to determine whether these people do in fact have these attributes. He says 'probably', indicating this was a subjective, fuzzy judgement based on gut instinct, not on careful 'evidence-based' study. Clearly, he used his intuition. 'Culture' as the object of analysis legitimates qualitative judgements like this.

For Widdows it captures the interrelations of assumptions, behaviours, values and outcomes. This complex whole controls his workforce, he says. If he can control what controls them, he has a powerful resource at his command. This is a resource available to him as a leader, not as a manager.

Neither he nor Day commented on it here, but there were good reasons why he could grasp the negative culture so rapidly. He had already encountered it in his own training as a manager. It is McGregor's 'Theory X'. Widdows' new system is a 'culture' based on McGregor's theory Y. As we saw in Chapter 4, all management textbooks and courses mention this theory.

Hitt et al. make this connection. They contrast Culture X, based on Theory X, with Culture Y, based on Theory Y (2007:118). Widdows' revolution was just a movement between two opposing sets of practices, both of which he knew. If we look at the story more closely we can see that Widdows used Theory X when he became CEO, when his new broom swept out 25% of the staff and restructured the Board. He was a master of Cultures X and Y, and never abandoned them. Even at the end of his revolution, he remained focused on the bottom line.

Culture and globalisation

The idea of 'culture' in the Organisation Culture tradition often neutralises the other main sense of 'culture' in business. Widdows, for instance, does not mention the diversity of his workforce or the cultural environment of the US company, Heinz, in Australia or in the Asia-Pacific region. In 1981, Japanese-American William Ouchi brought the theme of national culture into the mix when he made the US business best-seller lists with another letter: *Theory Z*.

Ouchi's echo of McGregor's work was deliberate, he said. His Theory Z had much in common with Theory Y, with one important difference. His aim, in the 1980s, was to explain to US business why Japanese management styles were so superior, and what US organisations could do about it.

His analysis began with McGregor's Theory X, so dominant in the USA that Ouchi called it Theory A. He labelled the Japanese style Theory J. The hope for change in the USA was the spontaneous discovery of Theory Y by some of the most successful firms. This made it possible for US business to learn and contribute to a new, more comprehensive theory. Ouchi's Theory Z was positioned off-shore, between the USA and Japan:

> Let us consider some maverick American companies that have many Japanese-like characteristics to pinpoint the ways in which we can indeed learn from the Japanese. (1981:9)

He calls these firms 'maverick', an American variant of 'larrikin', though the firms he mentions (e.g. IBM, Procter and Gamble, Hewlett-Packard) are not obviously mavericks or larrikins. He was urging US business to learn that it had to learn. At that moment the only cross-cultural source of knowledge was Japan, then out-performing US business spectacularly.

In a sense, we pursue a similar agenda. Theory Y, a subordinated tendency in business, can be developed in a richer cross-cultural field to produce a more comprehensive Theory Z. Contributing to this process we offer Theory L, Larrikin Theory. Theory L is fed from Theory Oz, Theory Mex, and Theory Braz. It includes lessons from difference and resistance, woven into a common theory and purpose.

Theory X is a top-down structure, imposed on workers understood to be reluctant and incapable, not to be trusted. For Theory X, larrikins serve no useful purpose. The 'culture' it expresses is the dominant culture of management, according to Smircich. Theory/culture Y trusts lower levels more, and empowers them. It is more larrikin-friendly.

Widdows initiated this change from above, but he also mobilised input from below. He invented the slogan, but the workforce were asked to give it content. Top management then put some of the proposals into practice. In this way Widdows partly follows the Larrikin Principle. He was a partial (fuzzy) larrikin.

Some important reflections arise from this analysis. In seeing the practices of 'Theory X' as a culture, he agrees with Smircich. If we see Theories X and Y and their relationship as all part of his culture, this will countermand the idea that a culture must definitionally be unitary. On the contrary, it is constituted around a contradiction.

Although it has the same form as the Managerialist Ideological Complex, it has a different origin and function. It is not just a form of discourse, designed to maintain the dominance of the dominant. It is a necessary set of practices that correct dysfunctions in the dominant culture and practice, which otherwise would destroy it.

The crucial problem is that methods of Managerialism, justified above all by claims to efficiency and productivity, produce the opposite. Widdows has stumbled on a form of the Management Governance Paradox. He also has a dim sense of what we see as the solution: the Larrikin Principle.

What is 'culture'?

An influential analysis of 'culture' across all its uses was written by Welsh-born Marxist literary theorist Raymond Williams, a Founding Father of the field of Cultural Studies. He identified 6 broad uses of 'culture' in English (1974). He showed that over its 2000-year history the word shifted and acquired an irreducible range of meanings. Yet 'culture' always identifies or creates a unity whatever it is in, and behind the diversity of meanings lies a unifying thread.

In our search for this unity and diversity we go back to Latin roots. *Cult-* came from Latin *col-*, which came from Indo-European *kwel-* to move around a place. 'Wheel' also comes from this Indo-European root. This gave *col-* a number of related meanings, depending on the object of the action.

These included land, hence to plough and cultivate land. 'Culture' in this sense refers to a kind of work or practice. It could refer to tending the body or mind, and hence to be 'cultivated'. From this came a connection of 'culture' with arts. It could refer to living in a place ('hang round'), from which 'culture' gained a connection with identity or belonging. Finally it could refer to worship, of the gods of the land. All these meanings still drew something common from *kwel*, the idea of repetition.

However dispersed and latent, all these meanings provide a background to the nuances and differences still present in modern 'culture', and related words in other European languages with the same ancestor (e.g. French *culture*, Spanish, Portuguese and Italian *cultura*, German *Kultur*).

Another word, not usually associated with *culture*, also descends from Latin *col-*: 'colony'. For Romans, colonies were places they laid claim to by living there, by cultivating them, by respecting the local gods, with the Roman army waiting in reserve in case these other reasons did not work. Colonisation was a soft word for soft power. European imperialism used 'colonisation' in similar ways.

'Culture' has long maintained close, if often hidden, connections with colonisation and power. As the primary object of study of Anthropology from the 19th century, it served the needs of European imperialism. Political and economic globalisation encountered many different, complex societies, driven by different assumptions, attitudes and sensitivities. 'Culture' became short-hand for what held these problematic factors together.

Palestinian-US writer Edward Said (1978) studied one strand of the global imperial project, colonisation of 'the East' (Asia and the Middle-East). He called the European cultural tradition devoted to the East 'Orientalism', a 'Western style for dominating, restructuring, and having authority over the Orient', 'a kind of Western projection onto and will to govern over the Orient' (1978:3, 95).

Said adapted Foucault's ideas of 'discourse' in order to describe the process. Orientalism, in his sense, is a set of discourses about culture, whose function is to manage a people through their culture. This makes 'Orientalism' a management device, applied on a massive scale, covering not just companies but nations and regions. It operates at the same fractal level as Neo-Liberalism, and has been incorporated into Neo-Liberalism's Ideological Complex, as we will see.

Like management ideology, Orientalism has a basically binary, power-driven structure, dividing the world into rulers and ruled, us and the others. 'Culture' in this process is an instrument of power, and a way of defining these others. Orientalism, according to Said, told those who lived in the Orient what they really were, and it told them that they could not otherwise know themselves.

We will give a simplified map of some key aspects of 'culture' in its modern senses and uses.

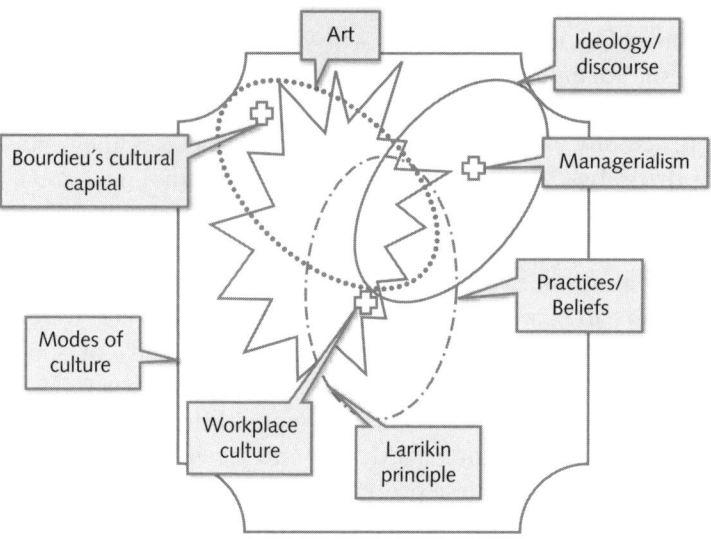

Modes of culture and the larrikin principle

This diagram shows the three main senses of 'culture' as art, ideology and practices and beliefs. Although the meanings are often distinct, they also overlap.

Within each sphere we show three-body systems as fractals of the larger complex. The ideological sphere of management has three focuses. To Managerialism and Neo-Liberalism, we add Orientalism, an older ideology that has not disappeared under 'modern' globalisation, just subsumed. Within culture-as-art, Culture as High Art coexists with popular forms. Bourdieu's idea of Culture as capital (1977), as a third body, is used for social discrimination. Within culture as practice, the three-body system includes practices and beliefs inside and outside the workplace.

The Larrikin Principle asymmetrically maps onto the three main tendencies in 'culture'. It arises in practice, inside and outside the workplace. It extends into artistic representations, in kinds of story and myth. It can be appropriated by ideology, but not as much as some other practices.

2. Ideological complexes and a multicultural world

'Culture' is a major presence in management writing, even more when dealing with 'cross-cultural' or 'global management' than when seeking to help managers like Widdows with organisational change. As the globalisation of business advanced, managers encountered problems and resistance to their plans, with 'culture' being brought into play to think about these difficulties. Its use in imperial anthropology prepared it for this role.

Culture and contradiction

Widdows saw the two cultures, X and Y, as homogeneous and mutually incompatible. He did not see that, in spite of their contradictions, the two co-exist in a single fissured culture to which he belongs. His assumption is shared by many who use the term culture, both classical anthropologists in the past and writers on management today. For classical anthropologists, a 'culture' was typically or definitionally small, primitive and cohesive

Said's Orientalism suggests one good political reason for this idea of culture. 'Orientalism' covered many different cultures, but the term 'orient' lumped them together, united only in being supposedly inferior to the 'culture' of the West. As a device for managing or despising peoples, it is easier to suppose they are all the same. This assumption maps readily onto 'Theory X' in Managerialism.

It was also easier to imagine them as suitable to be managed if these "inferior" cultures had no capacity for growth. If they could grow, they might change, perhaps into forms that challenged the inherent superiority of Western forms. As part of an ideology of management, they had to remain what they used to be.

Later critics called this idea 'essentialism', a term from philosophy. This idea supposes that cultures cannot evolve or change, or they will be untrue to their essential self. From our Larrikin Perspective, we notice that essentialists about culture do not read much philosophy, but absorb and reproduce ideological complexes.

Essentialism is criticised by most modern anthropologists and ignored by most business writers on culture. Indian writer Homi Bhabha (1994), like Argentinian-Mexican Nestor Canclini (1989), influentially promoted the term 'hybridity' to describe the complex state of cultures and identities in the contemporary globalising world.

The meaning is fine, but the biological metaphor is unfortunate. In biology, hybrids are sterile. The metaphor implies that these post-modern unions of different traditions are as unnatural as attempts to mate lions and tigers. They will not last. This attack on essentialism incorporates a version of essentialism.

We prefer another metaphor from biology, from distinguished US biologist Lynn Margulis. She developed the theory of *serial symbiogenesis* (1998) in an alternative theory of evolution. According to her, complex organisms developed as a result of the union of simpler cells into symbiotic forms. Four types of bacteria, combining in sequence, produced much of the incredible variety of the web of life.

According to Margulis, such fusions are more creative than fissions between species. In the process, the component organisms still survive within the new 'higher' form. Symbiogenesis, she says, is a powerful engine of evolution, yet tenaciously retains very ancient forms.

Mexican anthropologist Guillermo Bonfil (1987) independently developed an idea of cultural complexity which has similarities to Margulis's theory. The official ideology of Mexico proclaims its status as a nation of *mestizaje,* literally 'mixture', referring to the two broad streams, Indian and European, that make up most of its population. Bonfil saw this term as mere ideology, covering over a harsh truth. He captured it in a term that can be adapted to other nations with a history of colonisation. He coined the term *México Profundo,* Deep Mexico. Similarly we can talk of Deep Brazil and Deep Australia.

In contemporary Mexico, he said, the indigenous traditions of 'Mexico Profundo' were suppressed, 'a civilisation denied'. He asked:

> What does it mean in our history, for our present and above all for our future, this co-existence of two civilisations, the Mesoamerican and the Western? (1987:9)

He does not call them 'cultures' but *'proyectos civilizatorios'* (civilising projects), a term he developed with Brazilian theorist Darcy Ribeiro. We note his use of 'our'. The mixed identities are now a social fact, in himself and in Mexico. He insists that this is not a single fixed, unchanging culture, as in orientalist anthropology:

> It is not a passive, static world, but one which lives in permanent tension. The peoples of Mexico Profundo continuously create and recreate their culture, adjust it to changing pressures. (1987:11)

The two strands are not 'old' and 'new', 'primitive' and 'modern', but two ways of understanding human and social life. Each continued to evolve, and still co-exist, though one is denied and suppressed.

Bonfil's two 'civilizatory projects' do not correspond to Theories X and Y, or to Ouchi's Theory J or Theory Z. For him 'Mexico Profundo' evolved for many years as a project of the oppressed and despised indigenous people, a strategy for resistance. But it also had to be productive, a basis enabling a people to survive as well as they could in spite of injustice and oppression. Resistance alone was not enough.

In Brazil, a different civilizatory project encountered the same European pattern of colonisation, with African slave labour a crucial addition. At the ideological level there is the same celebration of multicultural diversity, and the same continuing systemic inequality. Again, one major product has been resistance, but there has also been creativity, mostly coming from below, from the interface between the two peoples, not from the supposed elite.

A similar pattern can be seen in Australia, again with differences. Australian Aborigines survived two centuries of ethnocidal assaults, continuing to constitute a Deep Australia. Differences between Anglos and Celts were smaller, but still significant. This is the double, double culture out of which the Australian Larrikin grew, further enriched with greater complexity and cultural productivity from later immigrant populations.

Margulis's term 'symbiogenesis' captures the productivity of the unions of peoples and traditions better than 'hybridity'. In this model the 'inferior' did not benefit from the admixture of qualities from the 'superior', as in imperialist ideology. Nor did the 'superior' lose their superior qualities by mixing them, as racist ideologies typically claim. Symbiogenesis can produce organisms that are more flexible, more creative and better adapted to the environment than either parent.

Within Margulis's biological model, we add Bonfil's premise. If one part of a tradition is repressed and denied, it will lose creativity and productivity, and the whole will suffer. In all three nations, deep culture is claimed and appropriated more by the rulers, but is lived, sustained and re-created more by the ruled. This is something for analysts of culture to always be aware of, as they listen to different versions of what is never the same culture, yet is still the same.

The managerialist ideological complex in cross-cultural management

The idea of culture in textbooks for cross-cultural, global or international management exemplifies a form of the management Ideological Complex. 'Culture' is a key contradiction in the complex, ameliorating harsher ideological forms of Managerialism and Neo-Liberalism, while attaching the older Orientalist complex.

We will illustrate how the complex works using the work of Helen Deresky, author of successful books on global management (1994, 1997, 2000, 2002, 2003, 2006). We use this single author because she allows us to show that these contradictions exist (in highly functional ways) within a single person, a single mind.

In Chapter 1 of *Global Management,* she describes 'The Global Manager's environment'. She gives a swift pen-portrait of the main blocks, and the features of the environment to attend to. One story is the triumph of Neo-Liberalism:

> One of the most striking changes today is that almost all nations have suddenly begun to develop decentralized, free-market systems in order to manage a global economy of intense competition, the complexity of high-tech industrialization, and an awakening hunger for freedom (2002:8).

She does not mention the managerialist methods that accompany the Neo-Liberal system, but this is in fact a 'major management problem' (2002:64) that the otherwise irresistible management methods keep encountering in global practice:

American managers' knowledge of other cultures lags far behind their understanding of other organizational processes (2002:8).

She quotes research that shows that 60% of expatriate managers leave their posts early because of poor performance, or they are ineffective. The losses to US firms amount to over USD 2 billion.

There is a contradiction here between the dominant ideological strand of Neo-Liberalism, as the One Best Way whose excellence is 'suddenly' (at last) being recognised as the universal best way, and a Theory X emphasis on 'fierce competition', drastic actions that competing organisations will have to take to survive. Theory Y celebrates 'the hunger for freedom' which will make the world a better, more united place.

But this ideological complex needs an Orientalist strand. Neo-Liberalism has no place for difference. There is only one way to succeed in business. No doubt is cast on these expatriates' 'understanding of other organisational processes', or on the universal applicability of these principles. It is just that there turns out to be something else to know about, to eliminate as a source of problems. As Nigel Holden writes, in a rare critical management text book on cross cultural management:

> [in traditional writing on culture and management] we can find that culture is about fundamental differences, which hang like swords of Damocles over international companies, poised to undermine their negotiations and pervert their strategies. (2002:xiv)

Yet culture sounds like a good thing, in Deresky and other management texts:

> As generally understood, the culture of a society comprises the shared values, understandings, assumptions and goals that are learned from earlier generations, imposed by present members of society, and passed on to succeeding generations. (Deresky 2002:64–65)

Values, understandings, assumptions and goals sound like good things, especially when 'shared' and 'learned'. But the other side of the ideological complex lurks in the shadows of this description. For instance, Deresky goes on to emphasise how this culture 'subconsciously guide[s] and control[s] certain norms of behaviours'. This sees 'culture' as an instrument of power and control.

Deresky's own 'values, understandings, assumptions and goals' come from her own culture, but she does not refer to this. They 'subconsciously guide and control' her norms and values too. This seems irrelevant to her. But they show up in subtle ways, in word choice or forms of expression. For instance, she defines culture as coming from earlier generations, so she assumes that it must be conservative and backward-looking. It is 'imposed', so it seems to be a kind of tyranny, autocratic, not democratic.

She calls it 'the culture', in the singular, implying that there can only be one culture, of 'a society', a whole nation moving as one, robots programmed by the culture planted in their subconscious. Since it is imposed by others it cannot be changed or negotiated. It is fixed in relation to the past and the present. Later generations who in turn have it imposed on them will be even more backward.

Given that this is the form of the Ideological Complex, we would be right to be suspicious of how the positive bits of the attitudes to culture and diversity really function in practice. We stay with Deresky. Here she offers some practical advice to would-be managers about bribes, a theme we dealt with in Chapter 7:

> They are called different names in different countries: tokens of appreciation, *la mordida* (the bite, in Mexico), *bastarella* ("little envelope" in Italy), *pot-de-vin* (jug of wine in France). For the sake of simplicity, we will categorize all these types of questionable payments as some form of bribery. In Mexico, for example, companies make monthly payments to the mail carriers, or their mail gets 'lost' (Deresky 2006:48).

We can see a number of features of Orientalism in action here. For instance, 'difference' is acknowledged everywhere, but only as different ways of doing the wrong thing. A package of non-English words is paraded and translated literally, showing how quaint and picturesque these people are in disguising the wrong thing. But 'we' cut through it all with the simple one-word judgement: 'bribery'.

This paragraph does not just inform students. It tries to construct them, shaping and modelling their perceptions and judgements. For instance, she puts inverted commas around 'lost', to imply that this is what 'they' say. 'We' (author and business students alike) are different. 'We' know how to decode it. 'We' laugh, with the Anglo writer, against the pathetic excuses of these corrupt Mexicans. 'We' not only learn the 'fact' about their corruption, 'we' also know whose side 'we' are on, what 'our' values are, and how universal they really are.

This curious strategy, showing seeming tolerance in order to reinforce existing prejudices, can be seen where Deresky is apparently recommending appreciation of the culture of the other:

> In Mexico, a visiting international executive or salesperson is advised to take time out, before negotiating business, to show appreciation for the surrounding architecture, which is prized by the Mexicans. (Deresky 2006:135)

This is advice to 'show' appreciation, not to feel it. This show is presented as a favour to these irritating Mexicans, who are made to sound like boring friends who insist on showing their holiday snaps.

This is a surprisingly patronising attitude to the great monuments and artefacts of Mesoamerican culture, 'culture' in another sense. American tourists flow across the border in their millions every year to visit this architecture, among the world's greatest tourist attractions. But Deresky seems not to consider the possibility that these Mexicans could have something interesting to say to visiting executives.

To do so would break the unwritten laws of Orientalism, that the Other cannot understand even their own culture without mediation by experts from the West (or North). In the discourse of tolerance, foreigners should pretend to respect the culture of the other, but never feel or allow genuine respect.

This advice relies on the simple natives to be so obtuse that they cannot recognize patronising ignorance. It will be bad advice, if they are not as obtuse as required. Deresky teaches a form of 'cross-cultural sensitivity' which displays arrogant insensitivity. It is designed to create goodwill in the natives, but is more likely to anger them, as it upset Gabriela when she read it.

Management's 'science' of culture

Deresky references her view of culture to Dutch-American theorist Geert Hofstede, whose *Consequences of culture* (1980) had a profound influence on the sub-field of management studies which deals with issues of cultural difference. Every management text in international business and cross-cultural or global management which we looked at had positive references to Hofstede. Many repeated his ideas without citation. He formed the taken-for-granted assumptions of management culture about 'culture'.

Hofstede was not an anthropologist. He came into management from engineering, his first field of study, in this respect like Taylor, father of scientific management. This mechanistic orientation, seeing humans as a kind of machine, can be seen in the way he framed the nature of culture: *'the collective programming of the mind that distinguishes the members of one group or category of people from another* (Hofstede and Hofstede 2005:4, their italics).

This is a Theory X formulation applied to a Theory Y theme. In this and other ways he seems a mediator between Theories X and Y. He urged Theory X managers to develop some interest (in certain circumstances) in culture and difference. But in doing so he behaved as a Theory X Orientalist within management. He explained what 'culture' means to anthropology and cultural studies with minimal reference to anthropologists, as if they do not understand their key term until someone from management explains it. *Cultures and organisations*, co-written with his son (2005), had only 10 references to classic anthropologists, about 3% of his total. Anthropology is not an equal partner in his construction of 'culture'.

Hofstede based his major work on a single study, a survey of 100,000 employees of IBM, across 72 countries. This basis gave him an immediate competitive edge in management studies over anthropologists, who typically had few numbers to back up their analysis. Yet as with all quantitative studies it is important to ask what is being counted, and how representative the samples are. Are employees from one MNC more valid sources of insight into culture than many voices and sources? His data shows what can be seen of other cultures from one window, a 'Western' MNC, whose biases distort just as much when multiplied 100,000 times

No overwhelming weight of empirical research makes his findings irrefutable. But most management writers are content to keep quoting his authority. We see this as a sign that he is producing something that management culture is so comfortable with that it sees no need to check it out.

Hofstede organised his findings around 4 main dimensions, which he called 'power distance' (PDI), 'individual-collectivist' (IDV), 'masculine-feminine' (MAS) and Uncertainty-Avoidance (UAI). Following are his figures for our three focal countries (from Hofstede and Hofstede 2005: 43–4, 78, 120–21, 168–69):

Country	PDI rank	IDV rank	MAS rank	UAI rank
Australia	61	2	20	55
Mexico	10	46	8	26
Brazil	26	38	37	31

The biggest problem with Hofstede's categories is the way they are applied in practice, as a grid which homogenises every nation they are applied to. Paradoxically Hofstede provides some of the data which could be used in this critique.

Like the ideological complex he influenced, Hofstede had a more complex view nested within his managerialist position. It is mostly buried in early parts of his texts, which managers who are keen to get to the useful part of the theory can be relied on to skip, but it is there. He acknowledged diversity within cultures. We all have

> several layers of mental programming within ourselves, corresponding to different levels of culture... [and] the mental programs from these various levels are not necessarily in harmony (2005:11).

He probed deeper, in an informal version of fractal analysis, when he considered different occupations within the same country. In these studies he found as much internal variation within countries as between them.

This implies that differences in class are as significant as differences in nationality. Struggles within nations and organisations produce cultural differences so important, even according to Hofstede's metric, that they should not be swept away, as is normally the case, by a single class-blind, difference-blind national category. Yet Hofstede's work is influential precisely because it ignores all this awareness with such authority, in a monumental study of diversity whose effect is to remove it.

Another example from Deresky shows how Managerialism creates an ideological amalgam from Hofstede and others. She reproduces a table from other authors comparing Japanese and Mexican cultures. We adapt it to include only the Mexican values. (Lawrence & Ryh-song 1994: From Deresky 2002:103).

Dimension	Mexican Culture
Hierarchical nature [PDI]	Rigid in all respects
Individualism v collectivism [IDV]	Collective relative to family group; don't transfer loyalty to workgroup; individualistic outside family
Attitudes towards work	Work is means of supporting self and family; leisure more important than work
Time orientation	Present orientation; time is imprecise; time commitments become desirable objectives
Approach to problem solving [UAI]	Reliance on intuition and emotion, individual approach
Fatalism [UAI]	Fatalism makes planning, disciplined routine unnatural
View of human nature [Theory X, Y]	Mixture of good and evil

We have indicated traces of Hofstede's categories. We have also annotated references to McGregor's Theories X and Y, which indicate how potent the half-life of this theory is in management ideology. Less clearly marked, but just as pervasive, is the hidden influence of Orientalism, which most managers have not heard of. All combine seamlessly to underpin the unstated, unshakable assumption that difference (from Anglo-American and, to a lesser extent, European norms) is inferiority.

For instance, the high PDI of Mexican culture is translated into the negative phrase 'rigid in all respects', as though Anglo-American MNCs do not have hierarchical structures either. The issue of time comes mainly from orientalist stereotypes. This table paints a picture of simple, primitive Mexicans who have not grasped the true nature of time in the business sense. Deresky deals with this theme in another place, where prejudice shows even more clearly:

> Polychronic people – Latin Americans, Arabs, and those from other collectivist cultures – may focus on several things at once, be highly distractible, and change plans often (2002:136).

In other contexts this could be seen as desirable. For instance, modern businesses require multi-tasking, multi-skilling, flexibility. But this connection is not made here. 'Polychronic' comes from the Greek, meaning 'many times'. This pretentious Greek word is used to repeat stereotypes about 'non-Western' peoples, including Latin Americans. They cannot concentrate, they cannot follow any discipline or routine, they cannot plan.

'Cross cultural understanding' in this form is simply a fancy way to despise Latin Americans and non-Western peoples with an easy conscience. It combines cultural prejudice and superiority while also feeling culturally sensitive. Hofstede's claims to be 'scientific' then give the final legitimation to this tissue of baseless irrationalities.

3. Cultural Analysis and the Larrikin Principle

In this final section of our book we want to translate Larrikin Principles into some practical tools for analysis, to help readers liberate what Audre Lord called the 'Master's tools' of cultural and social analysis for their own use. In the process we will emphasise the larrikin qualities of irreverence towards authority, flexibility of approach, and the war on bullshit. We set these attributes in the context of our awareness of the new scope and levels of chaos and complexity of the world in which the Larrikin Principle has to operate today.

Critical discourse analysis and ideological complexes

We have talked about critical discourse analysis as a good way of identifying strands and levels of ideological complexes. Here we give a concrete instance of such analysis as it deals with complexities of culture and ideology.

In May 2009 Mexican-US Sol Trujillo retired as CEO of Telstra, Australia's national telecommunications utility. He had been appointed by the Neo-Liberal government of John Howard to preside over a Neo-Liberal policy shift for the corporation, to partially privatise it and run it more efficiently. Trujillo resigned weeks before the due date, and gave a controversial interview.

He defended his record as CEO with two arguments. Australia is backward:

> I would say Australia is definitely different to the US. You know in many ways it's like stepping back in time just simply because of some of the policies, which are more recent. (Carswell 2009:2)

This has classic Neo-Liberal attitudes. 'Difference' (from USA) equals inferiority. Inferiority is read off a linear model of progress, in which later is definitionally better. Australian difference is seen, by this Mexican-American, as a negative, just as Mexican (and Brazilian) difference is judged in US management textbooks as inferiority. With these attitudes, Trujillo would be likely to offend the locals, as he did.

He also blamed his problems on racism:

> That does exist and it's got to change because the world is full of a lot of people and most economies have to take advantage – including Australia – of a diverse set of people. (Bingemann and Ooi 2009:3)

Trujillo cites one example of prejudice against him. Prime Minister Rudd, hearing the news of his departure, was reported to say 'Adios' (Bingemann and Ooi 2009:3).

The Australian media response to Trujillo was ferocious. Andrew Carswell, business editor of the *Daily Telegraph*, a populist/right wing newspaper in Rupert Murdoch's Newscorp stable, wrote under the headline: 'We're not racists, we just hate duds' (2009:2). He listed Trujillo's failures as CEO. Telstra's share price dropped from USD 5.05 at the time of his arrival to USD 3.15 at the time of his departure. He sacked more than 8,000 employees, but still pocketed over AUD 31 million in bonuses.

Such gaps between executive pay and performance were one facet of Managerialism that aroused most public anger in the wake of the financial crisis. Trujillo's case was typical. The share price dropped during the crash, but taking everything into account, Trujillo performed badly, for reasons unrelated to discrimination. He was a Neo-Liberal who mismanaged a semi-privatised entity, a manager who restructured while the profits and share prices were falling.

The Australian media attack had sound management reasons. At the same time it expressed larrikin outrage at managerialist bosses and practices. There was a further logical slide. Trujillo's failure as CEO did not prove his comments on racism wrong. On the contrary, on this point he was moderate, and right. Australia has undoubtedly been racist in the past (Hodge and O'Carroll 2006). In spite of recent improvements there is still a long way to go. The attack on Trujillo expresses Australian sensitivity. Australian racism also comes from below, from a current in Deep Australia. It has a place, contested but continuing, within the fissured, still growing culture. The angry, excessive denial of racism in Australia conceals and reflects these conflicts.

The comment he objected to from Prime Minister Rudd seemed so innocuous that to Australian readers it may seem a blatant over-reaction. Why should Rudd not say 'goodbye' in Trujillo's native Spanish, just as he impressed others talking Mandarin with the Chinese Prime Minister? But there is a huge difference in the two cases. 'Adios' projects an unstated racist stereotyping which ordinary Australians would pick up unerringly. So did Trujillo. He took offence, and was meant to.

In politics this strategy is known as 'dog whistling'. The term comes from whistles whose pitch can only be detected by dogs. It refers to racist or discriminatory comments, usually by politicians or shock-jocks. It works by triggering shared cultural assumptions. As in this case it makes provocative statements seem innocent, so that its target looks ridiculous for 'over-reacting'. Trujillo falls into this trap. The fact he does so shows he must have been sensitised to dog whistles in his past.

We are not defending Trujillo, but recognizing the complexity of his fuzzy identity. This explains more about him than crisp descriptions. As a US citizen with Mexican parents his fuzzy, layered identity includes something of the 'civilisation denied' that Bonfil mentioned, in proportions he unwisely did not factor in.

As a manager he displayed an excessive, counter-productive ferocity towards his political masters. For instance, in 2007 his corporation sued the government, the major shareholder, and the relevant minister, for awarding a contract to Telstra's competitor. This was larrikin disrespect for his boss and/or Neo-Liberal rejection of government interference, both to extremes.

He would have been a better manager if he had been more in touch with Deep Mexico. That would have allowed him to be more larrikin, more Australian. The paradoxes could have been used productively if they had been accepted and understood. In practice, in far-from-equilibrium conditions (a public utility in difficult transition to a Neo-Liberal order), they jeopardised the business and his own career.

This case illustrates how critical discourse analysis can go beyond ideological complexes mobilised in debates, supposedly under the control of senior management. There are deeper levels where ideology, culture and identity merge together, producing unpredictable responses and outcomes, for CEOs and larrikins alike.

Thick description

Classic anthropology provides complex terms for complex, non-linear patterns, and allows for some difference and diversity. These are advantages over the crisp linearity of management logic. For Widdows for instance, the term 'culture' became a signal for reflections on complexity and subjective dimensions.

But traditional essentialist orientations to culture have limitations. They assume that each system is self-consistent, self-contained and unchanging. Yet this is not the case with any complex cultural form, which is what managers and others always confront. With national cultures like Australian, Brazilian or Mexican (or US, French, Japanese etc.) the culture is so layered, fissured and contested that essentialist assumptions impose a distorting unity.

We want to capture shifting dynamics of far-from-equilibrium conditions, where the Larrikin Principle engages with the infinite complexity of global culture and processes. As a strategy for analysing culture and meaning in such fluid conditions, we value 'thick description', a term used by anthropologist Clifford Geertz (1973). Thick description is as multi-layered as the objects it describes. It can start from something as seemingly trivial as a wink or a smile or a phrase, with consequences as large and unpredictable as a butterfly effect. It sets its meanings, subtle traces of conflict and difference, in ever-wider frames of reference from local to global, like fractals.

Trujillo's reaction to Rudd's *'adios'* illustrates the diagnostic value of analysis informed by thick description. For Trujillo, the word in its context connected with many layers of history, personal, national and institutional. In the context it seemed an overreaction to some, but as with such incidents there were many deeper lessons to be learned. If we followed the implications through, it would end up in a three-body system including the USA, Mexico and Australia, and would not stop there. Analysis triggered by thick description can be inexhaustible.

As another illustration of the analytic power of thick description we take Widdows' smile, as photographed for his article. A smile may signify happiness, a man content with life who likes the person he is communicating with. His smile seems meant to say: 'A great place to work', reinforcing his slogan.

Yet an excess in this smile says 'this is hard work', and tension in his face and body emphasises the 'hard'. His shoulders are hunched, hands interleaved, holding something in. The messages express a contradiction which complements the contradiction of his partial larrikin pose. That combined power and egalitarianism. This expresses eagerness, vulnerability and stress.

We can follow these meanings up the fractal scale. Who is he smiling at? In the first place, the photographer, Lisa Said, whose name is in very small print. We know nothing about Lisa except what we deduce from her name. She is female, with a common Middle-Eastern name.

Through these bare details gender and cross-cultural dimensions, otherwise excluded from Widdows' heroic narrative, silently enter the scene, just outside the frame. It reminds us to understand Widdows' cultural change in wider terms than he mentioned, where changes he is not promoting may not even be on the agenda. But they could be. His organisation is situated in a cross-cultural environment, with the values and practices of women as well as men at issue, forces and factors not to be ignored.

Within Widdows' own story about his own change programme at his own company he can present a seamless, self-contained mix of Cultures X and Y under his control. As the story percolates up the various layers of a stratified world the coherence and stability disappear. The comforting fictions of managerial power seem less convincing against the uncontrollable world he chooses to ignore. The neatly managed contradictions of the managerial complex come apart, needing better, more sustainable solutions.

Our aim with this thick description is not anti-business. The delusions of the managerial complex are not, finally, in the interests of any manager. This analysis exposes realities he ignores at his peril. On a 'thin' reading, the article is close to propaganda, useless to anyone but the propagandist. Thick description makes it a rich source of lessons for everyone, at every level. This pseudo-larrikin CEO could learn greater respect for larrikin wisdom and all who carry it. If he did so, he would not be able to sustain most of the assumptions about 'good management' that he learnt in business schools or read in business journalism.

High art and the meaning of culture

'Culture' as high art seems to have drifted so far from 'culture-as-way–of-life' in societies and ideologies of organisations that it might as well be a different word. But high art (including painting and other plastic arts, literature and even film) reflects many important social meanings, relevant to students of organisations and cultures. We illustrate with Frida Kahlo, Mexico's internationally best-known female artist.

Her 'Two Fridas' (1939) combines two self-portraits, one clothed in a formal white dress, the other in the more simple, colourful clothes of a peasant. Kahlo's theme is the fissure in Bonfil's *Mexico Profundo,* which was still an issue for Trujillo 70 years later. Her point is unmistakeable, the sharp division along lines of class and ethnicity, which meet in the single (deeply divided) individual.

Both women are tightly constrained by social norms, indicated by elaborate clothes which reduce their movement. They hold hands formally, as though husband and wife in a conventional wedding portrait. Since both are women, this hints at lesbianism, even though girls holding hands was culturally acceptable in Frida's Mexico. Or since both are Frida this is auto-eroticism. This painting comes from a cultural insider who knows the core values well in order to challenge them.

The hearts are anatomical images super-imposed on the artistic representation. The scientific discourse of medicine jars with the discourse of art, yet is incorporated into it. This is a form of symbiogenesis, to use Margulis's term.

The style seems static, formal, even childlike, but its meanings are savagely subversive. Kahlo exploits a quality of art, especially high art, to create far-from-equilibrium worlds where critiques and connections can be more safely explored. This capacity to produce and see forbidden meanings in art is a powerful component of what Bourdieu (1984) called 'cultural capital'. An artist like Kahlo, who uses her freedom to shock a wide range of people, circulates that capital, making it grow.

Both Fridas communicate a sense of organisational structure that is rigid, constraining, and life-denying, as in Managerialism. But the third Frida, the artist, is very different. She was a rebel, breaking conventions in life, as in art. She was a kind of larrikina in her place and time. This larrikin passion, communicated by the painting though not directly represented in it, gives it its meaning and value. A work like this, analysed like this in a management course, could provide deeper insight into Mexico than the complete works of Hofstede and Deresky, plus the Mexican government website.

Cultural analysis and the larrikin myth

Earlier we introduced the idea of the larrikin as a myth. As our book ends, it is time to gather some of its threads together from this perspective. As others in critical management have recognised (e.g. Alvesson 2002, Czarniawska 1998 and Gabriel 2000), myth analysis can be a powerful diagnostic tool for organisation studies. In this final section we illustrate its flexibility and power for a synthesis of larrikin meanings.

We collected many stories from ordinary Australians, Brazilians and Mexicans. In most, larrikins were heroes, but in some, they were villains.

A balance like this is a common pattern. All three cultures produce both forms.

Outside our data we find a similar pattern in popular film. All 10 top box office Australian films had the Larrikin Principle as a main theme, and larrikin elements in its hero. Nine of the 10 top films of all time (all US-produced) similarly had a larrikin theme and hero. In the only apparent exception, *The sound of music*, Julie Andrews as a nun wears her habit with a difference.

Yet most management studies and textbooks ignore, marginalise or attack the larrikin theme. This may suggest that the culture of management is as different and unconnected from the surrounding culture as if it had existed on another planet. We take a different view. This duality exists within the two forms of culture, as well as between them.

Organisations of all kinds are open, dynamic systems at the edge of chaos. They direct flows, but they cannot stop them, or they die. Boundaries around organisations are so permeable and fuzzy that the respective cultures and myths must be seen as variants of a common base, just as Widdows' Cultures X and Y were opposing forms within a common culture.

As we indicated in Chapter 2, the myth analysis of Lévi-Strauss (1968) captures this complexity well. Very different myths for him can encode key categories of a culture, and explore its intrinsic anomalies and contradictions. He noted, of the 'trickster figure', whom we connect with larrikins, that

> the trickster is a mediator. Since his mediating function occupies a position halfway between two polar terms, he must retain something of that duality – namely an ambiguous and equivocal character (1968:226).

Gods likewise typically have contradictory attributes – the same god 'may be *good* and *bad* at the same time.' (1968:227). Myths have both heroes and villains, each equally illuminating. Widdows is a hero in his story, Trujillo comes across as villain. Yet both come from the same culture. They tell us about otherwise incommunicable problems of business and society, in Australia and the world.

Fuzzy logic helps translate Lévi-Strauss's ideas into analytic practice. The key terms in a larrikin lexicon are fuzzily both good and bad: the Brazilian *jeitinho*, Mexican *favorcitos*, larrikinism itself. The more we try to push these terms into a single consistent meaning, the more we lose their explanatory power for business and society, local to global.

'Thick description' helps keep this dynamic openness in analysis. For instance, Widdows' photo shows him as an uncomfortable part-larrikin. He is named as CEO, but not shown in a position of power, at his desk or at the head of a table. He crouches on the edge of an ordinary table. It is an asymmetrical double message: he is (mainly) formal and conservative, but is also (to some degree) informal, open to the views of others lower in the hierarchy.

He also mediates between Australian and International (US) cultures and identities. This is a major fault-line in business today. He was born and educated in Britain, and has worked globally, in Britain, the USA, Australia and New Zealand. But he 'solved' the wider cross-cultural problems by ignoring them. That is one way of mediating them, as ideology not myth or practice. Myth analysis would do a better job for him.

Trujillo was a failed mediator, between similar polarities. In his story, in the Australian media, the message was their intractability, and the cheeky non-prime-ministerial (larrikin) Rudd triumphed as hero. But this kind of analysis does not decide who is really the hero, or what happened. Rather, it brings out the deep contradictions of the categories and cultures in conflict.

As analysts of myth we suggest that Hofstede was influential because he combined soft (culture, anthropology) and hard (good 'theory X' management, quantitative analysis). Taylor's 'scientific management' was at the heart of Managerialism because it claimed to be scientific and was not, because it claimed to meet workers' needs for higher pay and better conditions as well as greater profits. Freidman's trick, for Neo-Liberal economics, was to claim that the unfettered pursuit of profit would produce more social goods than any intervention.

All four writers are regarded as theorists, but the success of their theories owes much to the myths they made. They show the continuing need for Tricksters in the world of business, even if (especially because) they did not claim to trick anyone. But that is the Trickster's typical final trick.

Precisely because it is so fissured and contradictory, the larrikin gaze illuminates core contradictions in management and business, especially in its leading doctrines and practices, its thinkers and heroes. The larrikin perspective tolerates and seeks out contradictions. It is more adequate the greater the range of cultures it accommodates. It is more stable and sustainable the more far-from-equilibrium the situation. It is what organisations most need but find hardest to accept. But that, like everything, could change. We hope our book may help.

Conclusion: A larrikin(a) critique

It is not in keeping with the Larrikin(a) Principle for us to finish with a neat, authoritative conclusion, crisply tying up all threads in a summary statement. Our theories say that would be bound to be limited and misleading. We mostly believe our theories. However, these theories do not reject all kinds of order or all kinds of synthesis. They pose a challenge: to articulate the different kind of order that emerges at the edge of chaos, producing fuzzy formulations which are complex, dynamic and above all useful.

We will organise some ideas around our three main themes and their intersections, The Larrikin(a) Principle, Managerialism and Neo-Liberalism as a three-body system.

Body 1: the Larrikin(a) Principle

The Larrikin(a) Principle has taken many forms under many names over its long history, as much in Mexico and Brazil as in Australia, Britain and the USA. Typically it has been carried by and within culture, though culture itself has its own contradictions and uses. Our concern in this book has been with the Larrikin(a) Principle as a local and global response to problems of organisational life that have become pressing over the past 50 years of international capitalism, made more visible during the global financial crisis of 2008-2009. We do not want to eliminate all alternatives to Larrikinism. On the contrary, it should be pervasive but co-exist with other different or even opposing qualities at every fractal level, a potentiality available to every nation, organisation and person.

Two strands make up the Larrikin(a) Principle, to different degrees, in all its incarnations. One is a distinctive attitude to formal rules and linear authority, affirming the values of informal systems, flexible practices, egalitarianism, loyalty to mates, and social justice. The other is a rejection of 'bullshit', especially as it emanates from authority figures and props up their exercise of power, damaging in the process the all-important dimensions of solidarity and trust.

We regard Managerialism and Neo-Liberalism as the two dominant systems that most need the Larrikin(a) Principle, though both these tendencies, like Larrikinism, have many names and histories, and take many different forms.

Body 2: Managerialism

Managerialism is our name for the dominant ideology of management in the contemporary capitalist world. It is distinct from the various best ways of managing that have evolved over different conditions, times and places, societies and cultures. On the contrary, it is an often dysfunctional set of prescriptions for management, which need to be bent or challenged by a version of the Larrikin(a) Principle.

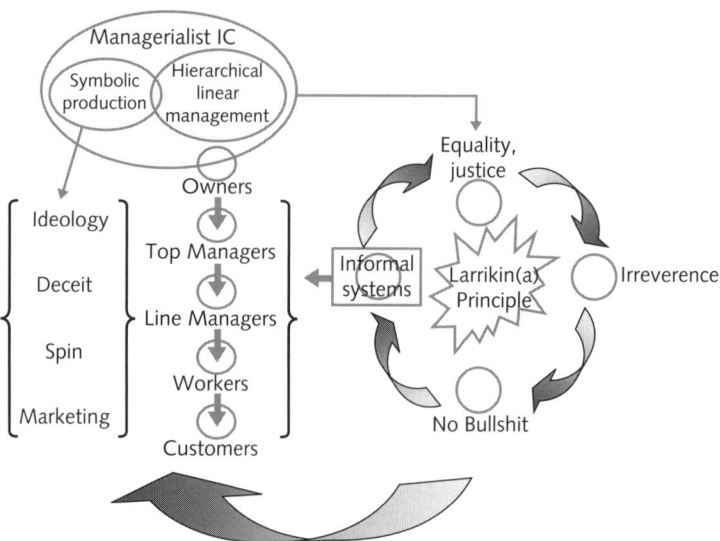

Managerialism is not a consistent picture of the world. It is an ideological complex, constituted by its contradictions, which constantly shift and change in response to the will and stratagems of the dominant. It operates through systemic deceptions to support the interests or illusions of the dominant. It needs the Larrikin(a) Principle's opposition to 'bullshit' to limit its tendency to create toxic environments inside and outside every organisation and every society. However, given the pervasiveness and tenacity of Managerialism and the dominant interests that it serves, we do not believe that the Larrikin(a) Principle or increased regulation are sufficient in themselves to overturn Managerialism.

Managerialism is not the same as management, even though its proponents try to blur the differences. The task and craft of managing are very different from Managerialism. In fact Managerialism produces typical forms of bad (ineffective, inhumane, unjust) management. The Larrikin(a) attack on Managerialism is not hostile to the function of management but conducted in the interests of doing it better. Managerialism is a dysfunctional set of prescriptions for management because it is based on hierarchical linearity. This constantly produces unintended, unproductive confusion and chaos, inefficiency and worse. Only the egalitarian, non-linear practices we associate with the Larrikin(a) Principle can deal adequately with the naturally-occurring complexity and chaos at every level of global life, and mitigate the damage created by the theories, policies and practices of Neo-Liberalism.

In our diagram we represent three main aspects of contemporary management theory and practice. In the centre is the dominant linear model, designed to enhance the power and profits of the class of owners alone. This is helped to survive thanks to the effects of two loops. On the right-hand side is the set of practices we call the Larrikin(a) system, which includes practices called by other names, such as informal systems and social capital. It counteracts extremes of linearity and injustice, and helps to make actual organisations and societies viable. On the left-hand side is the set of discursive and ideological forms which typically appropriate their content from the Larrikin(a) complex, though they are designed to preserve the central linear system of power and profit, and inhibit the growth and strength of the Larrikin(a) way.

This is also a strategic map to assist thought about the form a management revolution might take. Basic forms and formulae can already be glimpsed in the libratory policies and programmes already espoused in one wing of the Managerialist Ideological Complex (IC), in reformist ideas such as new paradigms, learning organisations and corporate social responsibility, already heard of by most people in most organisations, as aspirations if not reality. As Marx warned 150 years ago, any such revolution would have economic and political dimensions, which we do not try to diagram.

Third Body: Neo-Liberalism

Neo-Liberalism is our name for the dominant ideology framing the international strategies and operations of the economically-dominant nations, and the large corporations they support. It is experienced very differently from 'developed' nations like Australia and 'developing' or 'rapidly emerging' nations like Mexico and Brazil, and can only be really understood from both perspectives. Like Managerialism it is an ideological complex, constituted by contradictions, some of them between Neo-Liberalism and Managerialism itself. Like Managerialism but less directly, it needs a mix of Larrikin(a) qualities to moderate it: especially concern for justice and opposition to bullshit.

Like Managerialism, the fatal conceptual flaw of Neo-Liberalism is its linearity, its incapacity to understand the hyper-complexity of global processes which it claims to be the best system for mastering. The chaos and complexity dimension in the Larrikin(a) Principle is the best way of understanding and managing global hypercomplexity in the interests of a more just world. The Neo-Liberal system has a similar form to Managerialism, at a higher fractal level, with the same three strands. Its contradictory ideology masks a dysfunctional linear system of control, which is saved time and again by the complex and creative strategies of those it continues to exploit.

Managerialism in its global (Neo-Liberal) form celebrates a complexity and cultural diversity it accelerates and exacerbates but cannot understand or control. Transnational cultures are the medium for the negotiations, exchanges and transformations that are fundamental properties of globalisation in its fullest sense, as the medium for Larrikin(a) creativity. It is no accident that we have found so much resilience and creativity from below in Mexico, Brazil and Australia. They are testimony to the deep resources of wisdom and action that ordinary working people can draw upon in response to the structural injustices, dysfunctionalities, and vagaries of Managerialism and Neo-Liberalism.

The global financial crisis of 2008–9 did not introduce any real structural change. However, it exposed vulnerabilities and gaps in the dominant ideological complexes, briefly, if unevenly, producing a worldwide larrikin-friendly perspective. What it exposed was already known, more or less, by many prior to the crisis, and many after it have still not learned or accepted it. Its legacy will not so much be new institutional arrangements (likely to be weak and compromised, more regulated but not necessarily constraining the really powerful) but a new energy and scope for critique: empowered larrikins, supporting better ways of doing things, eroding the legitimacy and credibility of the rulers of the world.

Managerialism and Neo-Liberalism both operate parasitically on powerful, coherent alternative visions of the world that have evolved from below and are carried by the Larrikin(a) Principle. These alternative views of the world are used deceptively under Managerialism and Neo-Liberalism to limit and control real change. If they were uncoupled from their appropriation by managerialists and neoliberals, they could be a blueprint for a genuine, positive revolution. The Larrikin(a) Principle could play a role as a catalyst before and after such a revolution, if it happened. If it does not, the need for the Larrikin(a) Principle will be so much greater, just for sheer survival.

Bibliography

AEDT 2008 'Brazil announces record rate of Amazon deforestation' *ABC news,* http://www.abc.net.au/news/stories/2008/01/24/2146057.htm.
Althusser, L. 1971 *Lenin and philosophy* London, New Left Books.
Alvesson, M. and Wilmott, H. 1992 (eds) *Critical Management Studies,* London, Sage.
Alvesson, M. 2002 *Understanding organizational culture.* London, Thousand Oaks.
Amin, S. 1997 *Capitalism in the age of globalisation: The management of contemporary society* New York, Zed Books.
Anderson, C. and Kilduff, G. 2009 'Why do dominant personalities attain influence in face-to-face groups?' *Journal of Personality and Social Psychology* 96/2:491–503.
Anonymous 1947 Popol Vuh, (translator Recinos, A.), México, Fondo de Cultura Económica.
Ashby, W. R. 1956 *An introduction to cybernetics* London, Chapman and Hall.
Australian 2006 'Corruption scandal tightens Brazil poll' 3/10.
Australian Government 2007 *National statement on ethical conduct in human research* Canberra, Australian Government Printers.
Bak, P. 1996 *How nature works,* New York, Springer Verlag.
Bakan, J. 2004 *The corporation* Toronto, Viking Canada.
Ballinas, V. and Garduño R. 2008 'Detecta la ASF severas anomalías en el manejo presupuestal de Pemex en 2006', *La Jornada,* 28th March.
Banerjee, S. B. 2007 *The good, the bad and the ugly* Cheltenham, UK, Edward Elgar.
Barbosa, L. 1992 *O Jeitinho Brasileiro: A arte de ser mais mgual que os outros* (8th ed.). Rio de Janeiro, Editora Campos.
Barrow, C. 2003 *The handbook of business enterprise* Kogan Page India Private Limited.
Bartos, S. 2006 *Against the grain* Sydney, University of New South Wales Press.
Bateson, G. 1971 *Steps to an ecology of mind* London, Viking.
Bennis, W. and O'Toole, J. 2005 'How business schools lost their way' *Harvard Business Review,* May: 96–105.
Bhabha, H. 1994 *The location of culture* Routledge, New York.
Bijlsma-Frankema, K. And Costa, A. C. 2005 'Understanding the trust-control nexus', *International Sociology,* 20/3:259–282.
Bingemann, M. and Ooi, T. 2009 'Trujillo's "racist, backward" claims dismissed as sour grapes' *Australian,* 27/05:3.
Bonfil, G. 1987 *México profundo, una civilización negada* Mexico, SEP.
Bourdieu, P. 1977 *Outline of a theory of practice* (trans R Nice) Cambridge, Cambridge University Press.
– 1986 'The forms of capital' in J. Richardson (Ed) *Handbook of theory and research for the sociology of education,* New York Greenwood: 241–258.
– 2003 *Firing back* New York, New Press.
Braverman, H. 1974 *Labor and monopoly capital,* New York, Monthly Review Press.

Brazilista.com http://www.brazilista.com/ (Accessed January 5th 2004).

Broughton, P. D. 2008(a) 'Harvard loses its lustre' *Australian Financial Review,* 19 Sept.: 3.

– 2008 (b) *What they teach you at Harvard Business School* New York, Viking.

Browaeys, M.J. and Price, R. 2008 *Understanding cross-cultural management* London, Pearson Education.

Burrell, G. And Morgan, G. 1979 *Sociological paradigms and organizational analysis: Elements of the sociology of corporate life.* Portsmouth, NH, Heinemann.

Cabrera, A.; Cabrera, E. and Barajas, S. 2001 'The key role of organizational culture in a multi-system view of technology-driven change' *International Journal of Information Management* 21/3: 245–261.

Canclini, N. G. 1989 *Culturas híbridas* México, Grijalbo.

Carnegie, D. 1936 *How to win friends and influence people* London, Hutchinson.

Carswell, A. 2009 'We're not racist, we just hate duds' *Daily Telegraph* 27/05:3.

Castells, M. 2000 *The Rise of the network society* 2nd ed. Oxford, Blackwell.

Chomsky, N. 1999 *Profit over people* Seven Stories Press.

Clarke, T. and Clegg, S. 1998 *Changing paradigms* London, Harper Collins Business.

Clegg, S., Courpasson, D. and Phillips, N. 2006 *Power in organisations* London, Sage.

Clegg, S. 2008a 'Relationships of ownership, they whisper in the wings…' *Australian Review of Public Affairs,* March.

– 2008b 'There are no truths outside the Gates of Eden', Keynote speech at 2nd LAEMOS Colloquium Latin American European Meeting on Organizational Studies, Rio de Janeiro Brazil, 15–18 April.

Cohen, D. and Prusak, L. 2001 *In good company: how social capital makes organizations work* Boston, Harvard Business School Press.

Coleman, J. 1990 *Foundations of social theory.* London, Belknap Press/Harvard University Press.

Commonwealth of Australia 2006 *Corporate responsibility: managing roles and creating value* Canberra, Commonwealth of Australia.

Coronado, G. 2008 'Discourses of anti-corruption in Mexico: culture of corruption or corruption of culture?' *Portal* 5/1 Online, http//epress.lib.uts.edu.au/ojs/index.php/portal.

– 2009 'From autoethnography to the Quotidian Ethnographer: Analyzing organizations as hypertexts' *Qualitative Research Journal* 9/1:123–129.

Coronado, G. and Hodge, B. 2004 *El hipertexto multicultural en México postmoderno* Mexico, CIESAS/Porrua.

Coutts, L. 2004 'The resistance movement' *Business Review Weekly* Feb 19–25.

Covarrubias González, I. 2003, 'El impacto de la corrupción en el proceso de democratización de México. Una propuesta de análisis' *Revista Probidad* 24:1–14.

Czarniawska-Joerges, B. 1998 *A narrative approach in organization studies.* Thousand Oaks, California, Sage.

Czerniawska, F. and Potter, G. 1998 *Business in a virtual world,* London, MacMillan Business.

DaMatta, R. 1984 *O que faz o brasil, Brasil?* Rio de Janeiro, Brasil, Editora Rocco.

DaMatta, R. 1991 *Carnivals, rogues and heroes* London, University of Notre Dame Press.

Davidson, P. and Griffin, R 2003 *Management: Australia in a global context* Brisbane, John Wiley & Sons (2nd edition).
Davies, P. and Gribbin, J 1991 *The matter myth*, London, Viking.
Day, J. 2008 'Full of Beanz' *Management Today* 50, Nov/Dec.
Delves, P. 2008 'Lifting the bonnet on the world' *Spectator.co.uk* October 28th.
Deresky, H. 2002 *Global management. Strategic and interpersonal* New Jersey, Prentice Hall.
Deresky, H. 2006 *International management: managing across borders and cultures*, New Jersey, Pearson Education.
Douglas, M. 1970 *Purity and danger* Harmondsworth, Penguin.
Drucker, P. 1954 *The practice of management* London, Pan.
Drucker, P. 1993 *Post-capitalist society* Oxford, Butterworth-Heinemann.
Duarte, F. 2006 'Exploring the interpersonal transaction of the Brazilian *jeitinho* in bureaucratic contexts' *Organisation* 13/4:509–27.
Duarte, F. 2008 'What we learn today is how we behave tomorrow: A study on students' perceptions of ethics' *Social Responsibility Journal*, 1/2:120–128.
Duarte, F. and Hodge, B. 2007 'Crossing Paradigms: a Meta-Autoethnography of a Fieldwork Trip to Brazil' *Culture and Organization* 13/3:191–196.
Duarte, F., Gray, J. and McAllister, G. 2006 'Corporate Social Responsibility in a Dialectical Frame: the HIH Collapse in Australia' in D Crowther and K Caliyurt Eds. *Globalisation and Social Responsibility*, Cambridge, Cambridge Scholars Press.
Durkheim, E. 1952 *Suicide* London, Routledge.
Easterley, W. 2006 *White Man's burden* Harmondsworth, Penguin.
Ehrensal, K. 2001 'Training Capitalism's foot-soldiers: the hidden curriculum of undergraduate business education. In Margolis' *The Hidden Curriculum in Higher Education*, London, Routledge.
Foucault, M. 1971 Order of discourse *Social Science Information*, 10/2:7–30.
– 1978 *The history of sexuality: Vol. 1: An introduction* New York, Vintage Books.
– 1980 *Power/Knowledge* Brighton, Harvester.
Freeman, J. 1970 *The Tyranny of Structurelessness* http://flag.blackened.net/revolt/hist_texts/structurelessness.html.
Freeman, K. and MacNamara, A 2006 *In search of dark matter* New York, Praxis Publishing.
Freeman, R. 1984 *Strategic Management: a stakeholder approach* Boston, Mass., Pitman Publishers.
Freire, P. 1972 *Pedagogy of the oppressed* Harmondsworth, Penguin.
Friedman, M. and Friedman, R 1980 *Freedom to choose* Harmondsworth, Penguin.
Friedman, M. 1970 'The social responsibility of business is to increase its profits' Reprinted in L Hartman, ed., *Perspectives on business ethics* New York, McGraw-Hill Irwin.
Gabriel, Y. 2000 *Storytelling in organizations: facts, fictions, and fantasies*, Oxford, Oxford University Press.
– 2004 *Myths, organisations and structure* Oxford, Oxford University Press.
Garratt, B. 2002 'The learning organisation' in *Manager's Handbook*, ed. R Heller London, Dorling Kindersley.
Gates, B. 1999 *Business @ the speed of thought* Harmondsworth, Penguin.
Geertz, C. 1973 *The Interpretation of Culture* New York, Basic Books.

Gell-Mann, M. 1994 *The quark and the jaguar* London, Little and Brown.
Ghoshal, S. 2005 'Bad management theories are destroying good management practices' *Academy of Management Learning and Education* 4/1:75–91.
Gibbs, W. 2006 'Steve's memory will live on' *Women's Day,* 18/09:16–18.
Gimbutas, M. 1991 *The civilisation of the goddess* San Francisco, Harper.
Goleman, D. 1995 *Emotional intelligence* London, Bloomsbury.
Gome, A. and Ross, E. 2003 'The rising stars' *Business Review Weekly* Oct 16–22:36–47.
Grove, A. 1996 *Only the paranoid survive* New York, Currency Press.
Hardt, M. and Negri, A. 2000 *Empire* Harvard, Harvard University Press.
Harvey, D. 1989 *The condition of postmodernity* Oxford, Basil Blackwells.
Heller, R. (ed.) 2002 *Manager's handbook* London, Dorling Kindersley.
Heisenberg, W. 1989 *Physics and* philosophy, Harmondsworth, Penguin.
Hitt, M., Black, J. S, Porter, L. and Hanson, D. 2007 *Management* Sydney, Pearson.
Hobsbawm, E. 1969 *Bandits* London, Weidenfeld and Nicolson.
Hodge, B. and Coronado, G. 2006 'Mexico Inc? Discourse analysis and the triumph of managerialism' *Organization* 13/4: 529–547.
Hodge, B. 1978 *Foreshortened time* Cambridge, D S Brewer.
– 2008 'From other to self and back: the curious history of ethnography' in K Schlunke and N Sullivan, eds., *Cultural Theory in Everyday Practice.* Oxford, Oxford University Press.
Hodge, B. and Kress, G. 1988 *Social semiotics* London, Polity.
Hodge, B. and Louie, K .1998 *The politics of Chinese language and culture,* London, Routledge.
Hodge, B. and O'Carroll, J. 2006 *Borderwork in Multicultural Australia* Sydney, Allen and Unwin.
Hofstede, G. 1980 *Culture's consequences: international differences in work-related values* Beverly Hills, CA, Sage.
– 2001 *Culture's consequences: comparing values, behaviours, institutions and organisations across nations,* Beverly Hills, Sage.
Hofstede, G. and Hofstede, G. 2005 *Cultures and organisations* New York, McGraw Hill.
Hogan, P. 2006 'My mate the taxman' *Sydney Morning Herald* September 14th.
Holden, N. 2002 *Cross Cultural Management: A Knowledge management perspective,* UK, Prentice Hall.
Holton, J. 2007 *Insurgent citizenship: disjunctions of democracy and modernity in Brazil* New Jersey, Princeton University Press.
Hubbard, C. 2004 *Strategic management* Melbourne, Pearsons.
Kluger, J. 2009 'Why bosses tend to be blowhards' *Time* 2/03; 173/8:48.
Kostera, M. 2005 *Organisational epics and sagas* London, Palgrave.
Kotler, P., Haider D. and Rein I. 1993 *Marketing Places* New York, The Free Press.
Kuhn, T. S. 1962 *The structure of scientific revolutions* Chicago, Chicago University Press.
La Jornada 2008 'Energía, deuda y privatización' 24/03.
Lambsdorff, J.G. 2004 *Framework document: Background Paper to the 2004 Corruption Perception Index* Transparency International and University of Passau, Online http://www.ICGG.org?downloads/2004_CPI_FD.pdf. [Accessed 14 August 2005].
Lassere, P. 2003 *Global strategic management* Basingstoke, Palgrave Macmillan.

Lawrence, J.J. and Ryh-song, Yeh 1994 'The influence of Mexican culture on the use of Japanese Manufacturing techniques in Mexico' *Management International Review* 34/1:49–66.
Lévi-Strauss, C. 1963 *Structural anthropology* Harmondsworth, Penguin.
Loaeza, S. 2008 'Mala yerba...' *La Jornada* 3/04.
Lomnitz, C. (ed.) 2000 *Vicios públicos, virtudes privadas. La corrupción en México* México D.F., CIESAS/Porrúa.
Lorde, A. 1984 'The Master's tools can never dismantle the Master's house' *Sister/Outsider* Berkeley, Crossing Press.
Lorenz, E.1993 *The essence of chaos* Washington, University of Washington Press.
Lovelock, J. 1995 *Ages of Gaia* Oxford, Oxford University Press.
Mandelbrot, B. 1993 'The fractal geometry of nature' in Hall, N, ed. *The New Science Guide to chaos* Harmondsworth, Penguin.
Mann, M. 1999 'The dark side of democracy' *New Left Review* 235:18–45.
Marcus, G. 1995 'Ethnography of/in the world system' *Annual Review of Anthropology* 24:95–117.
Marginson, S. 2007 'Global setting, national policy and higher education in 2007' *Centre for the Study of Higher Education, University of Melbourne, A Public Seminar Series on Policy,* July 9th.
Marginson, S. 2008 'Sojourning students and creative cosmopolitans' in Peters, M, Murphy, P, & Marginson, S (eds) *Creativity and the Global Knowledge Economy* New York, Peter Lang.
Margulis, L. 1998 *Symbiotic planet: A new look at evolution* Basic Books, New York.
Marr, D. 2006 'Bosses illustrate fine art of belief in impossible things' *SMH* March 10th.
Marx, K. and Engels, F. 1970 *Selected works* London, Lawrence and Wishart.
– 1976 *The German Ideology* London, Lawrence and Wishart.
Maturana, H. and Varela, F. 1988 *The tree of knowledge* Boston, New Science Library.
Mauss, M. 1966 *The Gift* London, Routledge and Kegan Paul.
McGregor, D. 1960 *The human side of enterprise* New York, McGraw Hill.
McLean, B and Elkind, P. 2003 The smartest guys in the room Harmondsworth, Viking.
Menezes-Filho, N. and Vasconcellos, L. 2004 *Operationalising pro-poor growth* Washington, World Bank.
Merton, R. 1952 *Reader in bureaucracy* Glencoe, IL, Free Press.
Midalia, D. 2006 Letters, *The Australian* 8/09:15.
Mills, C. Wright 1959 *The sociological Imagination* Oxford, Oxford University Press.
Mintzberg, H. 1994 *Managers not MBAs: a hard look at soft practices of managing and management development* San Francisco, Berrett-Kohler.
Orwell, G. 1949 *1984* London, Secker and Warburg.
Ouchi, W. 1980 *Theory Z* New York: Avon Books.
Parker, M. 2002 *Against managerialism* Cambridge, Polity.
Parkinson, C. N. 1957 *Parkinson's law* Harmondsworth, Penguin.
Pascale, R .1990 *Managing on the edge* Harmondsworth, Penguin.
Peters, T. and Waterman, R. 1982 *In search of excellence* Australia, Random House.

Peters, T. 1987 Thriving on chaos New York, Alfred Knopf, Inc.
Porter, M. 2001 *The competitive advantage of nations* New York, the Free Press.
Prigogine, L. and Stengers, I. 1984 *Order out of chaos* New York, Bantam.
Pusey, M. 1991 Economic Rationalism in Canberra Sydney, Cambridge University Press.
Putnam, R. 1995 'Bowling alone: America's declining social capital' *Journal of Democracy*, 6/1: 65–78.
– 2000 Bowling Alone: The Collapse and revival of American community. New York, Simon and Schuster.
Quiggin, J. 1993 *Economic rationalism: dead end or way forward?* Sydney, Allen and Unwin.
Quiggin, J 2003 Blog: (johnquiggin.com.index.php/archives/2003/07/02).
Ricardo, D. 2007 (1817) *On the principles of political economy and Taxation* In Lewis, M ed., *The price of everything* New York, Sterling Publishing.
Robertson, R. 1993 *Globalisation: social theory and global culture* London, Sage.
Robbins, S. P. 2006 *Organization theory: Structures, designs, and applications* India, Dorling Kindersley.
Rodriguez, I. 2005 'Tiene Halliburton contratos con Pemex por más de mil 221 mdd' *La Jornada* 10/07.
– 2008 'Pese a ingresos históricos Pemex financia proyectos con deuda' *La Jornada* 25/05.
Rowley, T. 1997 'Moving beyond dyadic ties: a network theory of stakeholder influences' *Academy of Management Review* 22/4:887–910.
Rudd, K. 2009 'The global financial crisis' *The Monthly* No. 42: February http://www.themonthly.com.au/monthly-essays-kevin-rudd-global-financial-crisis--1421.
Said, E. 1978 *Orientalism* Harmondsworth, Penguin.
Samson, D. and Terziovski, M. 1999 'The relationship between total quality management practices and operational performances' *Journal of operational management* 17/4:393–409.
Sandoval, I. 2008 'Corrupción y petróleo' *La Jornada* 26/04:21.
Saunders, P, H. and Bradbury, B. 2008 *Poverty in Australia* Sydney, University of New South Wales.
Scharf, R. 2008 'Why Brazil leads the region in CSR' *American Quarterly* 2/1.
Schnews 2004 *Schnews at ten* Brighton, Calverts Press.
Schumacher, F. 1973 *Small is beautiful* London, Blond and Briggs.
Seligson, M. 2002 'The impact of corruption on regime legitimacy: A comparative study of four Latin American countries' *The Journal of Politics*, 64/2:408–433.
Smircich, L. 1983 'Concepts of culture and organizational analysis' *Administrative Science Quarterly*, 28:339–358.
Smith, A. 2007/ 1776 *An inquiry into the wealth of nations* In Lewis, M, ed., *The price of everything* New York, Sterling Publishing.
Soria, V. 1996 'Apertura económica, informalidad y Empobrecimiento en México' In *Alternancias y contradicciones del capitalismo* Mexico City, UAM Press.
Stiglitz, J. 2002 *Globalisation and its discontents* London, Penguin Books.
Subramanian, A. and Williamson, J. 2009 'The World crisis: reforming the international financial system' *Economic and political weekly* March 5th.

Sydney Morning Herald 2006 'OJ, me and my mate the taxman' by Paul Hogan, 14/09 http://www.smh.com.au/news/national/oj-me-and-my-mate-the-taxman-by-paul-hogan/2006/09/13/1157827020116.html.

Taylor, F. 1967 (1911) *Principles of scientific management* Easton, PA, Hive Books.

Thornton, M. 2005 'Universities; the governance trap and what to do about it' *Australian Fabian Society,* Melbourne March 16th.

Tracy, B. and Arden, R. 2006 *The power of charm: how to win anyone over in any situation* Amacom, American Management Association.

Triandis, H., Marin, G., Lisansky, J., and Betancourt, H. 1984 'Simpatia as a cultural script of Hispanics' *Journal of Personality and Social Psychology,* 47/6:1363–75.

Trinh, Minh-Ha 1989 *Women-Native-Other* Bloomington, Indiana University Press.

Tsoukas, H. 1999 'David and Goliath in the risk society' Organization 6/3: 499–528.

Utting, J. 2002 *Pragmatics and discourse: A resource book for students* London, Routledge.

Velez, C., de Barros, R, and Ferreira, F. 2004 *Inequality and economic development in Brazil* Washington, World Bank.

Weber, M. 1930 *The Protestant ethic and the spirit of capitalism* London, Unwin.

– 1947 *The theory of economic and social organisation* New York, Oxford University Press.

– 1948 *From Max Weber* Ed. H Gerth and C Wright Mills, London, Routledge.

White, B. 2008 'Then there were two very scared survivors' *Australian Financial Review* 19 September 26th.

White, M. 1997 *Narratives of therapists' lives* Adelaide, Dulwich Centre Publications.

Whorf, B. 1956 *Language, thought and reality* Cambridge, MIT Press.

Wiener, N. 1948 *Cybernetics* New York, John Wiley.

Wigand, J. 2007 *Testimony of Dr. Jeffrey Wigand before the House Subcommittee on Workforce Protection* 15/05. http://www.jeffreywigand.com/workforceprotections.php.

Williams, R. 1974 *Keywords* London, Collins.

Williamson, J. 1990 'What Washington means by policy reform' in J Williamson, ed., *Latin American Adjustments: how much has happened?* Washington, Peterson Institute.

– 2003 'From reform agenda to damaged brand' *Finance and development* September.

– 2008 The impact of the global financial crisis on Brazil, Washington, DC, Peterson Institute for International Economics.

Willmott, H. 1994 'Management education: provocations to a debate' *Management Learning* 25/1:105–36.

Zadeh, L. 1973 'Outline of a new approach to the analysis of complex systems and decision processes' *IEEE Transactions on systems, man, and cybernetics* 3/1:28–44.

Zipf, G. 1949 *Human behaviour and the principle of least effort* Cambridge, MA, Addison-Wesley.

Subject index

Australia 13–14, 25, 26, 27, 28, 33, 36, 37, 52, 56, 58–61, 66, 67–9, 70–1, 73, 74, 80, 83, 84, 105, 108–10, 118–9, 131, 133, 135, 139–141, 150–1, 163, 175, 181, 186, 187, 196, 198, 203, 206, 209, 214, 219, 221–4, 226–9, 231–2

Brazil 13, 14, 24, 25, 26, 27, 28, 29, 35, 41–2, 56, 59, 67, 69, 73, 106, 115, 123, 131, 132, 135, 139–40, 141, 142, 147–8, 151, 153, 155–6, 157–9, 162–6, 170, 173, 175, 181–3, 185–8, 195, 198, 203, 214, 219, 222, 223, 226, 227, 229, 231–2

Butterfly Effect 21, 22, 91–2, 95, 224

Change 12, 15, 19, 23, 42, 45–6, 57, 74, 75, 79, 81, 86, 95–100, 116, 117, 122, 191, 204, 205, 209–10, 213, 224

Chaos theory 20, 22–4, 38, 40, 49, 54, 62, 65, 88, 112, 117, 149, 161, 168, 177, 179, 192–3

Charm, *charme* 181, 182, 184–7, 192, 195, 196, 200

Colonialism (post-, Neo-) 42, 70, 115

Control 11, 21, 22, 23, 32, 36, 37, 38, 43–4, 48, 72, 76, 78, 85, 87, 93, 94, 96–7, 104, 109, 116, 134, 146, 151, 163, 166–7, 168–71, 172, 177, 180, 194, 202, 216, 223, 225, 232

Corporate Social Responsibility 173, 175, 231

Corruption 28, 76, 130–1, 132–3, 134–56, 170, 196, 203, 206, 217

Critical Discourse Analysis 221, 223

Critical Management Studies 15, 16, 35, 173, 204, 205

Critical pedagogy 37, 41, 46, 49, 102

Crocodile Dundee 131, 132

Cross-cultural management 23, 215

Cybernetics 22–3, 71–3, 86–7, 101, 121–2, 145, 174

Dark matter 54–5, 56, 57, 58, 62, 74, 76, 78, 93, 118, 156, 191

De-skilling 34

Dysfunctional Organisation 145, 172

Economic Rationalism 52, 83, 108, 163, 164, 165, 190–1

Far-from-Equilibrium Conditions 63, 76, 79, 88, 97, 123, 124, 130, 144, 148, 160, 170–1, 187, 223, 224

Favorcito 186, 192, 227

Fordism 87, 89, 116

Fractals 23–4, 43, 112, 212, 224

Fuzzy (logic) 23, 24, 26, 61–2, 72, 227

Gender 13, 26, 28, 57, 58, 60–1, 66, 72–3, 185, 224

Global Financial Crisis 10, 21, 103, 229, 232

Globalisation 12, 18, 21, 25, 28, 41, 52, 59, 66, 69, 75, 86, 104, 114–7, 118–24, 127, 128–9, 159, 170, 178, 192, 209, 211, 212, 232

Globality 159–60, 161, 177

Governance 10, 137, 142, 156, 201, 208, 210

Hidden curriculum 35

Hydra 103, 106

Ideological complex 19, 24, 32, 36, 39, 41, 50, 73, 77, 79, 87, 89, 90–1, 94, 100, 102, 103, 104, 108, 110–4, 118, 124, 128, 140, 156, 162, 163, 164, 173, 174, 188–9, 191, 197, 199, 208, 210, 211, 212–3, 215–7, 219, 221, 223, 230, 231, 232

Informal Systems/ Polity 14, 55, 58, 93, 131, 134, 179–80, 181, 182, 183, 184, 186, 188, 194, 197–201, 203, 206, 229, 231

Invisible hand 21, 112, 114, 115, 159, 177, 182, 185, 202

Jeitinho 133–4, 187–8, 192, 227

Larrikin/Larrikina 13–28, 29, 32–3, 35–7, 38, 40, 41, 42, 44, 46, 48–9, 50, 51, 53, 54–6, 58–78, 79–82, 83, 84, 85, 87, 89, 91, 92, 93–4, 95, 96, 98–100, 102, 106–7, 108, 110, 111, 117, 118, 119, 121, 123, 125, 127, 128, 130–2, 133–4, 136, 138–9, 143, 145, 150, 155, 156, 157, 158, 159, 161, 163, 166, 168, 169, 175, 177, 178, 179–82, 185–8, 191, 194, 197, 199, 201, 203, 204, 207, 208, 209–10, 212, 213, 215, 221–3, 224, 225, 226–8, 229–32

Latin America 26, 35, 42, 105, 107, 111, 112, 132, 141, 153, 156, 184, 221
Learning organisations 86, 87, 93, 231
Linearity 20, 28, 31, 87, 95, 197, 208, 223, 231
Management education 33, 34–5, 37, 38–9, 40, 41, 42–3, 44, 49, 50, 52, 135
Managerialism 11, 13, 18, 19, 20, 21, 22, 23, 28, 32, 50, 54–5, 73, 75, 79–86, 91, 93, 94, 95, 98, 102, 103, 104, 114, 116, 117, 118, 127, 128, 130, 134, 144, 146, 149, 157, 163, 165, 168, 174–5, 179, 180, 190, 191, 197, 203, 204, 207, 210, 212, 213, 215, 220, 222, 226, 228, 229, 230–2
Marcos, Subcomandante 70
Meta-communication 101, 146
Mexico 24, 25, 27, 28, 35, 42, 43, 55, 56, 58–9, 67, 69–70, 73, 104, 106–7, 113, 115, 118–20, 125, 131, 135, 137–8, 140, 141–4, 151–6, 157, 165, 175, 181, 182, 198, 201–3, 214, 217, 219–20, 222–4, 225–6, 229, 231–2
Mordida 142, 217
Multi-national Corporations 113, 119
Myth 28, 51, 65–7, 68, 72, 90, 95, 96–7, 100, 102, 125, 212, 226–8
Ned Kelly 68, 90, 92, 115
Neo-Liberalism 10–13, 15, 18, 19, 21, 22, 23–4, 28, 32, 46, 52–3, 70, 83, 103–13, 114, 123, 124, 127–9, 130, 134, 135, 146, 147, 151, 153, 156, 157, 159, 160, 161–2, 163, 164, 165, 171, 174, 182, 189, 190, 191, 202, 203, 211, 212, 215, 216, 229, 231–2
Networks 28, 77, 124, 170, 172, 175–6, 180, 182, 184, 188, 190, 194, 196, 197–200
Non-linearity 20, 178, 192, 208
Paradigm 41, 79, 84, 86–9, 93–5, 101, 116, 118, 164, 231
Power 19, 24, 38, 43, 62, 66, 67, 76, 77, 78, 79, 82, 88, 92, 94, 96, 98, 100, 102, 110–1, 112, 113, 115, 125, 131, 136–7, 144, 146, 147, 148, 149, 150, 151, 153, 154, 157, 160, 162, 164–6, 168–71, 174, 177, 184, 185–7, 189, 193, 203, 211, 216, 219, 224, 225, 226, 227, 229, 231
Restructuring 52, 80, 82–3, 96, 114, 163, 211

Retrovirus 80–1, 87, 90, 92, 93, 97, 116, 154, 188
Schizophrenic/schizogenic 89, 105, 145, 146, 147, 174
Semiotics 15, 49, 187
Simpatia 182, 184–7
Social Capital 28, 55, 172, 179–80, 181, 183–4, 188–9, 190, 191–2, 194–7, 199, 203, 231
Soft Capital 179, 182, 194, 203
Soft Systems 180, 182
Space-time compression 122–3
Stakeholder model 173–4
Star Wars 94
Steve Irwin 60
Taylorism 87, 88, 89–90, 91, 92–3, 96, 97, 101, 116, 118
Thick description 223–5, 227
TQM 62, 80, 81, 85, 87–8, 90
Trust 11, 14, 44, 51, 57–8, 77, 84, 91, 95, 132, 136, 166–70, 178, 181, 190–1, 194, 209–10, 229
USA 10, 12, 20, 25, 32, 35, 42, 59, 68–9, 72–3, 89, 92, 113, 131, 135, 136, 141, 152, 185, 189, 196, 203, 209, 221–2, 224, 228, 229
Washington Consensus 105, 106, 107, 111, 112, 113, 124, 135, 148, 160, 162, 201
Whistleblowers 76, 85–6, 135–6

Author index

Althusser, L 38–40, 43–4
Alvesson, M 15, 205, 226
Bakan, J. 94
Banerjee, S. B. 173–5
Barbosa, L. 188
Bateson, G. 101–2, 145–6, 149
Bennis, W. and O'Toole, J. 30–2, 35, 41, 46, 49, 81
Bhabha, H. 213
Bijlsma-Frankema, K. and Costa, A. C. 166
Bonfil, G. 214–5, 223, 226
Bourdieu, P. 179, 188–9, 191, 194, 197, 212, 226
Braverman, H. 85
Burrell, G. and Morgan, G. 205
Canclini, N. 213
Carnegie, D. 185
Castells, M. 117
Chomsky, N. 106, 111, 124
Clegg, S. 9, 16, 29, 32, 36, 41, 52, 79, 86–7, 95, 116, 118, 168
Cohen, D. and Prusak, L. 188, 190–1
Coleman, J. 188–91
Czarniawska-Joerges, B. 65
Czerniawska, F. and Potter, G. 226
DaMatta, R. 157, 186–7
Davidson, P. and Griffin R 37, 40, 73–5, 98–9, 114, 207–8
Deresky, H. 117, 215–8, 220, 226
Douglas, M. 66, 93
Drucker, P. 116
Duarte, F. 51, 100, 158, 188
Durkheim, E. 169–70
Easterley, W. 18
Ehrensal, K. 35, 38–40
Foucault, M. 89, 96–7, 168–71, 211
Freeman, J. 54, 127, 174
Freeman, K. and MacNamara, A. 54
Freire, P. 41–4, 46, 49, 175
Friedman, M. 52, 111–2, 125, 147–9, 153, 174
Gabriel, Y. 65, 126
Garratt, B. 86–7, 93, 177
Gates, B. 20, 117
Geertz, C. 224
Gell-Mann, M. 21–2, 24

Ghoshal, S. 32, 45–6
Gimbutas, M. 66
Grove, A. 101
Hardt, M. and Negri, A. 115
Harvey, D. 89, 118, 122–3
Heisenberg, W. 122–3
Hodge, B. 11, 18–9, 44, 51, 67, 100, 119, 121, 125, 222
Hofstede, G. 205, 207, 218–21, 226, 228
Kuhn, T. S. 79
Lévi-Strauss, C. 65–7, 72, 77, 227
Lomnitz, C. 155–6
Lorde, A. 18
Lorenz, E. 21, 91, 160
Lovelock, J. 121–3
Mandelbrot, B. 23–4
Marginson, S. 52–3
Margulis, L. 213–5, 226
Marr, D. 150, 151
Marx, K. and Engels, F. 19, 41, 86, 114–6, 148, 191–3, 195–6, 231
Maturana, H. and Varela, F. 155
Mauss, M. 56, 190
McGregor, D. 93, 208–9, 220
Merton, R. 165
Mintzberg, H. 30, 35, 41, 46, 49
Orwell, G. 38, 43
Ouchi, W. 209, 214
Parker, M. 16, 35, 38
Parkinson, C. N. 74–5, 84
Pascale, R. 79
Peters, T. 20, 205, 207
Porter, M. 141
Prigogine, L. and Stengers, I. 20–2, 121, 123, 170–1, 192–3
Pusey, M. 52, 83, 108, 163
Putnam, R. 179, 188–9, 191, 194, 199
Quiggin, J. 82, 84, 97, 108
Ricardo, D. 88, 191
Robertson, R. 120, 159
Said, E. 18, 211, 213
Schnews 125–8
Schumacher, F. 18
Smircich, L. 205, 210
Smith, A. 21, 111–2, 115, 159, 182, 185, 191–2, 195, 202
Stiglitz, J. 52, 106, 112–3, 124, 160–1

242

Taylor, F. 89–98, 101, 111, 116–8, 191, 218, 228
Tsoukas, H. 125
Weber, M. 87, 123, 164–5, 208
White, M. 92, 193
Wiener, N. 22–3, 160
Wigand, J. 76–8, 106, 136
Williams, R. 100, 210
Williamson, J. 76, 105–6, 108, 112, 124, 135, 160, 165
Willmott, H. 41
Zadeh, L. 23–4, 40, 61–2, 65, 72, 88, 95, 109, 123, 144, 167
Zipf, G. 24